Materials and Thermodynamics

Materials and Thermodynamics

Living and Economic Systems

Pierre Delhaes

iSTE

WILEY

First published 2017 in Great Britain and the United States by ISTE Ltd and John Wiley & Sons, Inc.

ISTE Ltd
27-37 St George's Road
London SW19 4EU
UK

www.iste.co.uk

John Wiley & Sons, Inc.
111 River Street
Hoboken, NJ 07030
USA

www.wiley.com

Library of Congress Control Number: 2017945468

British Library Cataloguing-in-Publication Data
A CIP record for this book is available from the British Library
ISBN 978-1-78630-208-3

Contents

Preface

The initial aim is to classify materials from generalized thermodynamics outside the equilibrium state and not according to their chemical origin. For this, a thermodynamic system is defined according to its environment and its compliance. An ideal isolated system will be defined by the thermodynamic functions at or close to equilibrium. However, we must consider a real system that may be closed, that exchanges only energy and may contain data, or, an open system when there is also an exchange of matter. The level of these exchanges or flows, assuming there are sufficient reserves, makes it possible to distinguish between a linear regime for the response of the system or outside of it. In the latter case, these are non-linear dissipative systems with the appearance of new organizations, spatio-temporal structures, which can go as far as a state of deterministic chaos. The chemical reactions or the materials and their arrangements demonstrate the validity of this approach, which can be extended to biology and living systems. This description is accompanied by a description at the microscopic level in statistical physics that allows the introduction of quantum mechanics and that of information theory. Biomimicry is a stumbling block to analyze the behavior of nano and biomaterials in this classification linking microscopic and macroscopic aspects. The most innovative domains appear to be those of molecular electronics, bionics and quantum computing.

This classification can be extended to living, ecological, economic and financial systems with similar behaviors: all these systems can be classified according to their deviation from an ideal situation of thermodynamic equilibrium. The concepts of dynamic complexity and hierarchy, emphasizing the crucial role played by cycles and rhythms, then become fundamental. Finally, the limitations of the uniqueness of this description, that depend on thermodynamic foundations based on concepts of energy and entropy, are discussed in relation to the cognitive sciences.

Pierre DELHAES
June 2017

Introduction

We use more and more objects, products of human creation. They are made of a particular material with a specific shape that gives them functionality for the desired use. These objects become ever more elaborate and they form ever more sophisticated assemblies or devices up to the design of machines. To describe and classify these objects, the usual method is to observe their behavior and analyze the phenomena generated. To do so, an overall scientific approach is needed that will allow the modeling of these behaviors by choosing the most general approach possible whatever their chemical origin. Physical models are generally reductive concepts of reality, initially based on the existence of mechanical systems. However, the transition from mechanics to phenomenological thermodynamics, while retaining the variational principles of stability on the extrema of potential functions, is the product of conceptual advances made over the past two centuries. The energetic approach with the introduction of the temperature variable (T) and associated quantities has become indispensable. By doing so, the principle of energy conservation and that of evolution from the entropy function become unavoidable. We will show that a thermodynamic approach allows a transversal analysis with a more general classification than that based on the chemical nature of the materials. These criteria, which are developed for isolated thermodynamic systems, in or close to equilibrium, are only valid for an ideal system. However, in practice, a system exchanges energy and ultimately matter with the exterior: it should be far from equilibrium. A dissipative behavior occurs, which we will introduce. This phenomenological description can then be extended to living environments and to economic sciences with an increasing degree of complexity; we will discuss this in the three last chapters.

Finally, this approach at the macroscopic level will have to be accompanied by a microscopic description involving the achievements of statistical physics and the principles of quantum mechanics. The theory of information will also be included in this microscopic description, including the experimental methods of storage and reading of computer data. As the dimensions of the devices only decrease over time, new classes of nanomaterials are introduced, including those of biomaterials and their mimetic approach. They therefore require a quantum description of physical phenomena. It is then necessary to reconcile these two levels of approaches related to the problem of irreversibility of time and the principle of evolution in a thermodynamic system far from equilibrium. For this reason, we have divided the text into three main parts:

1) Classical phenomenological approach and functional classification of materials:

– Chapter 1 recalls the historical character based on the relationship between the material and the shape of the object or device; these respective approaches are schematically attributed to Aristotle and Plato. Increasing knowledge of materials shows how geometry, particularly that of polyhedra, has influenced the development of chemistry and biology and allows the introduction of materials;

– Chapter 2 is an overview of useful phenomenological thermodynamic definitions and different possible situations for an isolated thermodynamic system, then exchanging with its environment. They vary in distance from equilibrium and are classified using the instability threshold concept and the appearance of new organizations. They are called spatiotemporal structures that can go as far as chaotic situations. Thus, in this context, the analysis of the processes for the production of solid-state materials helps justify the forms characterized at different scales and obtained by the processes of cristallogenesis or morphogenesis;

– the classification of materials proposed in Chapter 3 is a consequence of this thermodynamic concept. It is based on the intrinsic properties of a condensed phase organized by adding the active role of the surfaces and interfaces, that is to say, the control of possible exchanges between the system and its environment. This approach makes it possible to functionally classify the main families of experimental devices for given constraints and flows by assessing the observed responses;

– Chapter 4, a logical continuation of the previous one, focuses on their use dedicated to the energy capture and storage and the acquisition of information. It presents materials and their assemblies that can convert and store electrical energy, vital secondary energy. Then, the recording, storage and reading of information using different types of digital memories is discussed by comparing the different technical solutions proposed.

2) Microscopic and nanomaterial description:

– Chapter 5 marks a breakthrough in this approach with the introduction of microscopic models and statistical thermodynamics. The analysis in terms of a particle assembly comes down to processing in statistical physics and the assumptions of quantum mechanics. This approach makes it possible to redefine the thermodynamic quantities, in particular the concept of statistical entropy. The information theory is then introduced within the framework of this generalized microscopic description;

– Chapter 6 discusses the experimental component by introducing, on the one hand, synthetic chemistry, the creation of new functional materials, and, on the other hand, the development of the main nanotechnologies needed to develop them. These are the two branches of a step toward accessing the nanoscience;

– Chapter 7 is a continuation of the previous chapter with developments related to engineering and molecular electronics. In fact, in this extreme case, the system consists of only a few hundred or tens of entities and even possibly a single molecule. We predominantly present the concrete results in the fields of nanoelectronics, nanophotonics, also nanomagnetism. They lead to the development of quantum memories and their own logic for a computer of the future.

3) Extensions to living organisms and economic systems:

– Chapter 8 makes it possible to move onto living organisms by first recalling their behavior at the molecular and cellular level. According to the same energy approach, they are systems outside the equilibrium leading to the creation of a specific organization in living environments. The aim is to discuss work on biomaterials and biosystems, which are often at the nanoscale and achieved by mimicry of these organisms. The problems with energy conversion and storage will be addressed as well as those concerning genetic computing and biotechnologies;

– Chapter 9 is devoted to the application of generalized thermodynamic concepts for living species, their populations in ecological, geological and natural environments. They show spatiotemporal structures that can reach a chaotic state. In particular, the behavior of hierarchical systems with natural or anthropomorphic rhythms and cycles is discussed;

– In Chapter 10, a thermodynamic extrapolation of the macroeconomy is introduced. Considered as an experimental science based on energy, the contribution in terms of thermodynamics out of equilibrium is analyzed. Then, we examine the dynamic aspect of economic and, in particular, financial systems, which can go as far as unstable situations of a chaotic nature;

– finally, a summary in the form of a conclusion discusses the classification of materials beginning with the introductory history of the form and the functionality. The analysis of the thermodynamic, energy and entropy concepts of increasingly complex systems is also revisited. In these dynamic systems, the concepts of hierarchy, cycles and rhythms are finally analyzed. To conclude, the validity and limitations of this generalized thermodynamic approach is discussed.

In this horizontal analysis, we will not build on the fundamental properties of matter but instead their technological development: how to conceive and create an object or "artifact" or a device with a shape optimized for the chosen use. The aim is to classify the thermodynamic responses of such a system as a function of constraints, external fluxes or even punctual stimuli. The keywords that are frequently used are defined in a glossary to clarify their meaning in this context. With regard to the bibliography, it lists reference books or general articles that help provide supporting information on the topic tackled. They are often supplemented by recent publications that illustrate the state of progress of a given subject but are not exhaustive, especially in the final chapters.

On the conceptual level, as the chapters proceed, we shall see that a whole stream of thought has been developing for almost a century, materializing with the help of scientists who could have been potential Nobel Prize winners such as Shannon, Jaynes, Brillouin, Landauer or Mandelbrot. Indeed, all these real thermodynamic systems, whose theories were established during the 20th Century, operate outside the ideal state of equilibrium. They are dynamic, conditioned by their exchanges with the environment, energetic mass or even data. Initially located near equilibrium, they show a linear response that becomes insufficient as the distance from equilibrium increases. From a critical point or threshold, they give rise to dissipative nonlinear systems that can present novel self-organizations and possess a deterministic behavior defined as chaotic. This is what we will discuss by first focusing on the behavior of the different classes of materials.

1

Form and Matter: The Genesis of Materials

Prehistoric man discovered tools with polished and cut stone, the first steps in the use of natural resources. The history of humanity can be characterized by the nature of the objects used: ages of stone, bronze, iron, carbon (coal), silicon (or polymers), etc. Since then, technical advances have only increased and multiplied by developing the idea of substance and then matter associated with the progress of knowledge. In this context, a historical approach to the concept of materials is essentially linked to the development of the physical sciences since antiquity. Historians place their birth in Greece several centuries B.C. where it was only a part of knowledge grouped under the name "philosophy of nature" [ROS 79]. The associated classical image is that of the school of Athens with its two most famous representatives, Plato and Aristotle painted by Raphael (Figure 1.1). A simple approach relates the complementarity between shape and matter in the creation of an object. A typical example is that of a statue in which the sculptor imposes a particular form deemed ideal, with materials coming from different origins (marble, bronze, wood, etc.).

Since antiquity, this complementarity has persisted; this is what we will show in the initial part of this book. Therefore, we will pinpoint the key chronological events by distinguishing the different geometric forms involved and their evolution, independently of their content. The Renaissance in Europe is considered to be a fundamental period, followed by the establishment of the modern foundations of chemistry and physics from the 17th and 18th Centuries. Then, the creation of synthetic materials, characteristic of contemporary chemistry, sparked their extraordinary evolution.

In the second part, we will show how these forms have become active surfaces or interfaces associated with a deeper description of matter always present in a limited volume. Thus, the notion of a finite system and of exchanges with the exterior becomes concrete when defining an object and its functionality. This approach is in fact a description and an analysis through the prism of our current knowledge where generalized thermodynamics will play an essential role.

Figure 1.1. *Central part of Raphael's painting (c. 1520) entitled "the School of Athens" exhibited at the Vatican museum in Rome. Plato on the left points his finger to the sky while Aristotle on the right has his hands directed toward Earth: these gestures symbolize their respective philosophical approach of knowledge, rather idealist or even realistic. For a color version of the figure, see www.iste.co.uk/delhaes/materials.zip*

1.1. Role and evolution of geometric shapes in chemistry

1.1.1. *Shape and matter: the origins*

When historians analyze the birth of Greek science, they recognize two main contributors. Following Thales, the school of Pythagoras developed mathematics, in particular, geometry, and Empedocles proposed the existence of four primordial elements. In Athens, Plato adopted the system based on these four principles: fire, air, water and earth, associating a mathematical concept [BAU 04]. In the Timaeus, Plato proposed a geometric model where these elements are represented by regular polyhedra as shown in Figure 1.2. These five regular polyhedra consist of identical faces of variable numbers; triangular faces for the tetrahedron (fire), octahedron (air), icosahedron (water) and square (earth) but also pentagonal faces for the dodecahedron, representing the universe. In this context, we should mention the mechanistic approach of Democritus, who proposed the existence of a smaller piece of indivisible substance called a particle, a concept that we shall find much later with the existence of the atom.

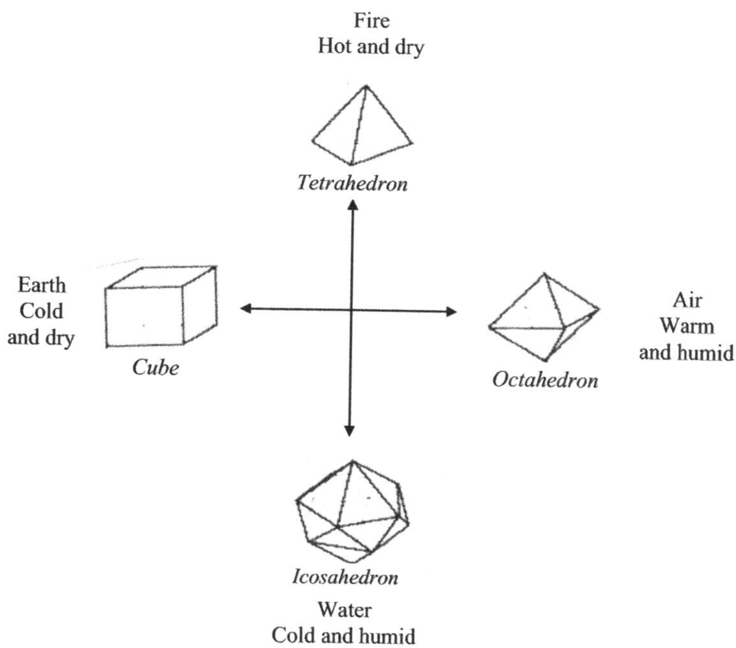

Figure 1.2. *The four primitive elements identified as Plato's regular polyhedra and associated with the qualities described by Aristotle (adapted from [LAU 01])*

The next step is due to "Aristotle's Physics" that attributes great importance to the observation and development of a method of reasoning. It is a substantialist approach that attributes an essential role to the perceptive qualities of man (the warmth, the cold, the wet, the dry). To change one element to another, one must act on a single quality or a couple of them [LAU 01], which is shown in Figure 1.2. Thus, Aristotle constructed a system by establishing a logic based on the symbiosis of shape and matter (called hylemorphism). Besides the results, establishing a scientific method through research into the causes is the main advantage. The exploration of materials mainly took place in Alexandria with the birth of Alchemy, which would be resumed and pursued by the Arab civilization. With regard to the gradual transition of natural substances to those isolated by laboratory methods, we can cite the eighth Century work by Gerber, who purified metals and various salts [BER 85].

1.1.2. *From the Renaissance to modern chemistry*

At the time of the Renaissance in Europe, toward the end of the 15th Century, the ideas of Plato and Aristotle were revived and idealized. Raphael's painting (Figure 1.1) symbolizes this appropriation. It shows Plato who seeks an ideal by raising his finger to the sky to suggest the notion of mathematical abstraction, and Aristotle who looks at the Earth and nature in general, sources of matter. Ideas of geometric space and perspectives will come through painting. They are associated with the rediscovery of the regular polyhedra drawn by Leonardo da Vinci in a work by Luca Pacioli ("De divina proportiona" published in Venice in 1509). Indeed, it was artist-engineers like Leonardo da Vinci or Albrecht Durer who brought a renewed interest at the scientific level. A significant example is an engraving called "Melancolia" where Durer drew symbols around the character, including a balance, an hourglass, a magic square, a sphere and an irregular polyhedron in which a face is reflected (see Figure 1.3). A little later Kepler embedded Plato's solids into a sphere, which is used as a model of the solar system ("Mysterium Cosmographicum" published in 1596).

The beginning of the 16th Century is also characterized by the revival of alchemical works that isolate pure bodies and medicine with the preparation of medicines: a typical approach is that of Paracelsus and his successors. The dominant fact is the birth of modern science toward the middle of the century and during the next century, reinforced by the rationalistic approach of Descartes. The creation and development of instruments for observation and measurement, especially in optics, have been decisive steps since the time of Aristotle.

Figure 1.3. *Copper engraving by Albrecht Durer (1513) famous for its elaborate composition showing a set of symbolic objects*

During the 18th Century, the importance of polyhedra was still significant. Theoretically, Euler established general geometrical relations to construct regular or semiregular polyhedra. Experimentally, the mineralogical contributions of various scientists, such as Haüy, a contemporary of Lavoisier, showed the presence of natural polyhedral crystals that appeared to be fundamental for geometrical crystallography (see the image presented in Figure 1.4). These polyhedra represent a concrete reality that will continue to develop.

From this revolution, modern chemistry was gradually born through the establishment of a nomenclature and new symbols to classify matter [LAU 01]. Many steps had to be taken over two centuries; these are described in the chapter on chemistry in the third volume of the encyclopedia of Sciences, Arts and Crafts edited by Le Rond d'Alembert and Diderot in 1780. The birth of chemical elements was initiated by Lavoisier followed by an advance in the structure of matter associated with the definition of a chemical system, proposed later by Dalton. The concept of an atom and then of a molecule, going back to the idea of Democritus, was debated throughout the 19th Century: it was formalized with the establishment of the periodic classification of elements by Mendeleev in 1879 [ROS 79]. Let us conclude by indicating that the 19th Century was characterized by the study of chemical transformations and the creation of synthetic products, especially in organic chemistry. Its extension to the living domain with the birth of biochemistry and the establishment of a generalized chemical language were the main achievements.

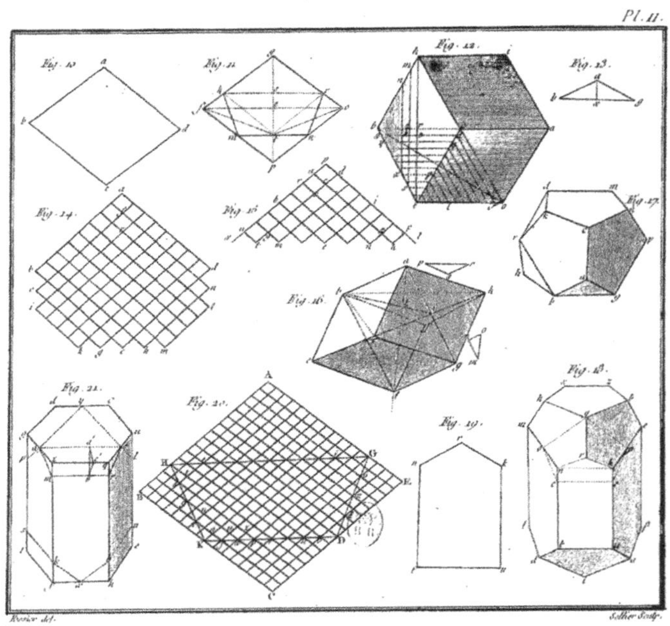

Figure 1.4. *Example of an image showing various polyhedrons drawn by Haüy. "Essay on a theory of the structure of crystals applied to several kinds of crystallized substances" (editors Gogué and Wée de la Rochelle, Paris 1784)*

1.1.3. *Modern era*

The beginning of the 20th Century was characterized by an in-depth study of the constitution of matter at the microscopic level because of two essential discoveries: X-ray diffraction and observations with electron microscopy [GUI 80]. It was around 1912–1915 that Le Von Laue and Bragg, father and son, discovered the method of diffraction of an electromagnetic wave allowing the reconstruction of the atomic structure of crystallized solids. Then in the 1930s, Siemens engineers in Germany designed microscopes to visualize the texture of these solids, followed by scanning electron microscopy and then electron diffraction microscopy. Various instruments have been developed to produce high-resolution microscopes that are able to “see” the atoms. In this technological context, the discovery of near-field microscopes was made during the 1980s, allowing a more topographical description, that is to say, the visualization of surfaces. Even more recently, the development of fluorescence optical microscopes has led to significant advances in biology. Thus, these observation techniques are derived from the matter–radiation interaction and depend on the depth of penetration of the matter by the electromagnetic wave involved. They allow characterizations at several scales schematically from the micron to nanometer scale and are qualified as textural and structural, respectively.

These technical advances throughout the 20th Century have led to the return of geometric forms in chemistry both atomically and molecularly. The seven crystallographic structures listed are often the images of regular forms observed in mineralogy (see Figure 1.4). They obey particular rules, essentially the conditions of translation invariance and symmetry operations that define these crystal lattices. In particular, they exclude axes of symmetry of orders 5 or 7 in a plane, which do not lead to a complete filling of the Euclidean space. Work on non-crystallized solids such as glass has led to changes in these concepts. We must first mention the discovery of quasi-crystals in which a symmetry axis of order 5 is present: the local structure is a pentagon that does not lead to satisfactory planar tiling. This situation is to be compared to the theoretical tiling of Penrose but it is possible to form a two-dimensional network by bending the plane that eliminates possible hindrances. Indeed, 12 pentagons are combined in a sphere giving a dodecahedron. In this way, we find the last regular polyhedron identified by the Greeks, which can form a topological space that is curved rather than Euclidian, as proposed by Poincaré at the beginning of the 20th Century. By bending the space, the regularity of the crystal structures is seen despite appearing disordered, as in the case of glass [SAD 92].

At the level of the molecular structure, stable aggregates and cage-shaped solids can be also obtained by forming more elaborate polyhedra. Figure 1.5 shows two recent examples of monoatomic assemblies, composed of carbon and silicon, respectively. The discovery of fullerene in 1985, a molecule comprising 60 carbon atoms, was an epistemological breakthrough in chemistry. This molecule is an icosahedron, a semiregular polyhedron known since Archimedes and described by Pacioli. It is the best understood molecule of this family, which consists of closed cages with a variable number of carbon atoms. With regard to silicon, we also see in Figure 1.5, a stable assembly of 33 atoms, which is a particular example of an aggregate. Variable polyhedral cages exist by forming clathrate-like phases. These situations show the extent of the polymorphism observed for different solids [DEL 11a].

An extension of these simple examples involves porous three-dimensional structures: these are zeolites of natural or synthetic origin [FER 07]. A historical example is shown in Figure 1.6 by presenting the atomic structure of Stilbite, a zeolite that is found in its natural state. This figure shows the presence of atomic tetrahedra, which play a fundamental role in modern chemistry. At present, a large number of alveolar solids have been synthesized by manipulating their architecture, size and geometry of the channels and access windows. This geometric parameter makes it possible to control their porosity, an essential parameter in surface chemistry.

Ultimately, these examples show how structural techniques have contributed to the better understanding of atomic and molecular structures, which are often constructed from polyhedra introduced by Greek philosophers (see Figure 1.2). In doing so, we have gone from a macroscopic description to a microscopic analysis of matter thanks to technological improvements, leading to a gradual reduction in the observation scale.

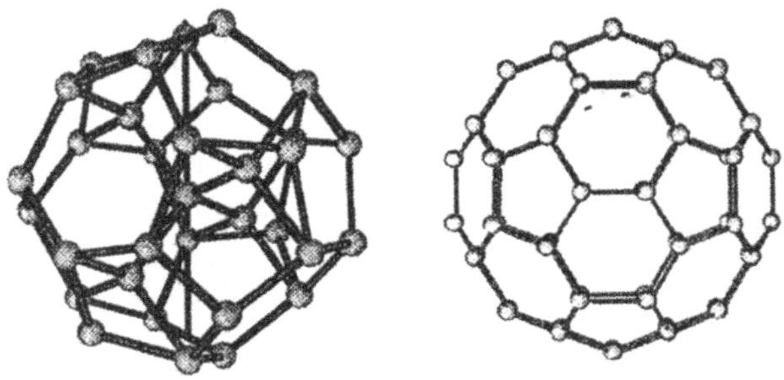

Figure 1.5. *Examples of cage structures: on the right, fullerene C_{60} formed of pentagons isolated by hexagons, and on the left, a silicon aggregate consisting of a polyhedron with 28 atoms and a tetrahedron Si_5 inside*

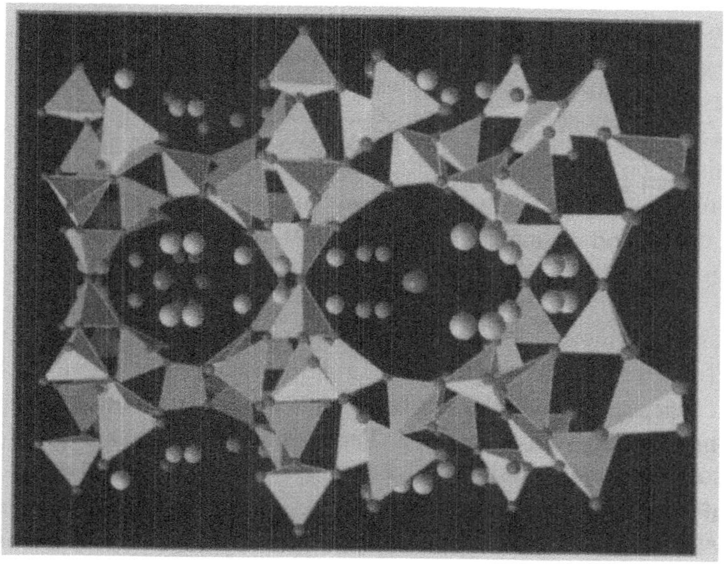

Figure 1.6. *Structure of Stilbite, the first porous solid discovered in the 18th Century; the skeleton is formed by the combination of silicate and aluminate tetrahedra generating tunnels in which sodium (gray) and calcium (blue) ions are found (according to [FER 07])*

1.2. Contributions of complexity of forms and thermodynamics

1.2.1. *Development of more complex forms*

The emergence of geometric forms more complicated than those of polyhedra led to the description of inert assemblies being extended to living environments. The founding works are attributed to D'Arcy Thomson in his book "On growth and form" published in Cambridge in 1917. It developed this field of research toward morphogenesis and the study of the mechanisms of formation of condensed matter. In this part, although we are interested in the inert matter observed under different conditions, the transition to a living environment is a critical aspect, which has initiated many studies. Thus, in an evolutionary medium subjected to exchanges, the appearance of more and more complex forms is observed. Various growths then occur at the atomic or molecular scale, defined as crystallogenesis but also at higher scales known as mesoscopic or macroscopic by morphogenesis. The purpose of understanding the formation and stability of these forms in the living world is to exploit them by biomimicry to create shapes, and ultimately, related functions. A well-known example is the creation of chemical gardens following work initiated by Leduc ("The Physical Basis of Life and Biogenesis", Masson, Paris 1906), which attempted to recreate mineral growths resembling living organisms by means of an osmotic effect [EAS 09]. This approach based on the self-formation of organized sets and generation of controlled chemical forms has recently been reviewed [BAR 15]. An example of macroscopic concretions confined between two plates is shown in Figure 1.7.

Systematic kinetic studies have been undertaken to analyze the forms obtained [COL 08]. A significant example is shown in Figure 1.8 where different types of crystallization of calcium carbonate are obtained according to varying experimental conditions. These mesocrystals result from different kinetics to the formation of polyhedra due to the involvement of weak chemical interactions. Thus, various forms are obtained by self-organization: they are analogous with biominerals, suggesting the idea of a continuous process.

Figure 1.7. *Variety of chemical garden obtained by the addition of cobalt, copper, iron, nickel and zinc salts in a solution of sodium silicate (according to Barge et al. [BAR 15])*

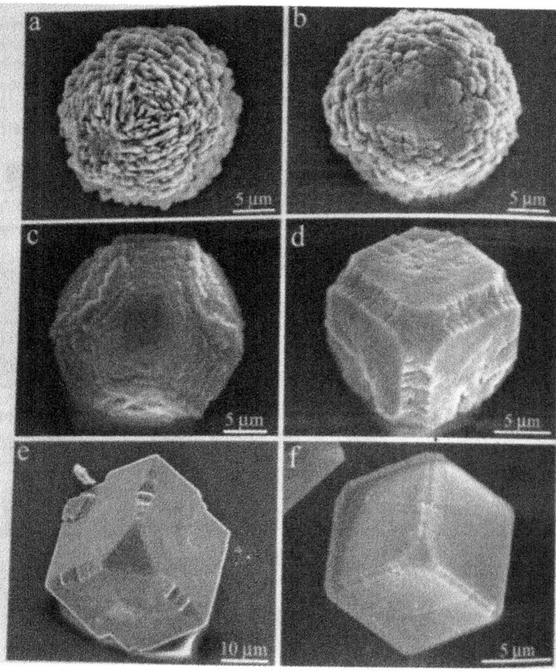

Figure 1.8. *Photographs of calcium carbonate crystals precipitated from an aqueous solution containing a constant molar ratio of calcium and sulfur and with variable concentrations decreasing from plate (a) to (f) (according to [COL 08])*

Recent studies confirm that, under the influence of a variable environment over time, different growth regimes exist and "floral" hierarchical structures are obtained and controlled [NOO 13]. Finally, the geometric extension to living forms considering developmental biology and involving morphogenesis cell mechanisms has been recently developed [PRO 08]. Particular examples include leopard spots, zebra stripes or even the formation of polygons in giraffes' skin, resembling Penrose tiling. This point will be discussed in Chapter 2 describing the morphogenesis of materials, then in Chapter 9 for living systems.

1.2.2. *Introduction to generalized thermodynamics*

As we have just discussed, the emergence of evolutionary forms has been the subject of a large number of studies over the last century. Two main groups are represented by either a thermodynamic [GLA 73] or a mechanistic [THO 72] approach. To illustrate this, we have shown in Figure 1.9 the covers of two books typical of these approaches and published almost simultaneously around 1970. For Thom [THO 72], it is a dynamic analysis of spatial structures and their transition from the catastrophe theory, which is considered as a neo-Platonic topological description on the geometry of shapes. On the contrary, for Glansdorff and Prigogine [GLA 71], the generalization of the concept of thermodynamic systems far from a state of equilibrium is fundamental. The stability or otherwise of the system depends on the nature, intensity and control of its exchanges or flows with the exterior, starting with the energy. Indeed, the significant contribution made over the past century has been the extension of the bases of phenomenological thermodynamics to situations outside an equilibrium state in a finite system interacting with its environment. The classic concepts of an isolated system and associated state functions and then their generalization for concrete situations in the presence of exchanges will be presented in Chapter 2.

Thus, these two schools of thought have existed since antiquity; the second more abstract one is centered on the mathematical description of dynamic phenomena that are often unstable, whereas the first one is more materialistic and directed toward modeling energy in a given situation. This is the second approach that we will develop.

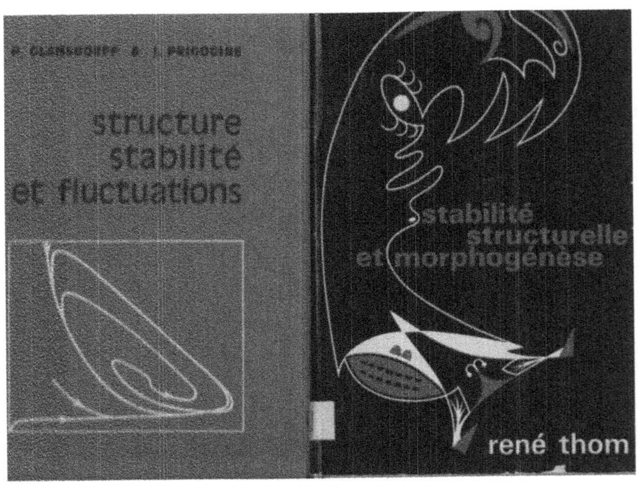

Figure 1.9. *Covers of books, published simultaneously after 1970, on the stability and evolution of dynamic systems: to the right that of Thom [THO 72] emphasizing the geometry of the forms, and to the left, written by Glansdorff and Prigogine [GLA 71], where thermodynamic fluctuations play an essential role*

1.2.3. *Toward a classification of materials*

A generalist approach amounts to considering thermodynamics as the general basis for the study of materials. Therefore, they are like finite systems in interaction with their environment. The creation of an object for a specific use will depend not only on the intrinsic volumetric properties but also on the forms and characteristics of the interface. The nature of energy and/or mass exchanges will determine the field of application and a material can be defined by the physicochemical functionality of its surface or interface. This purpose leads to the existence of various classes of increasingly sophisticated materials that have followed the evolution of science and technology over the centuries.

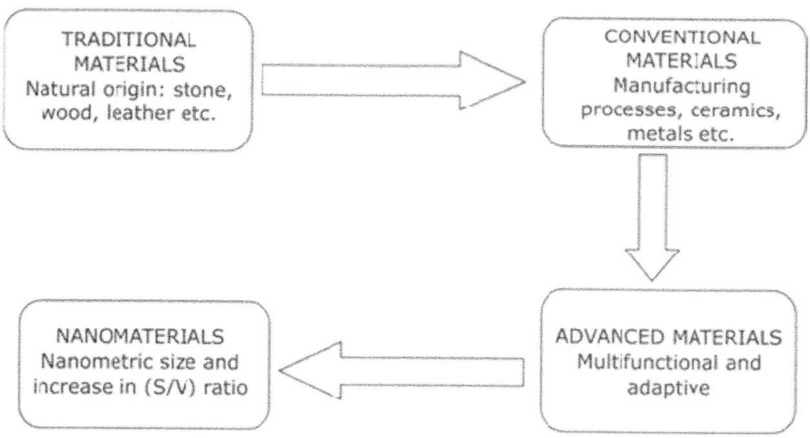

Figure 1.10. *Chronological approach to the evolution of materials*

A diagrammatic representation of the evolution of the main classes of materials is presented in Figure 1.10: it shows the major technological developments since early history with the birth of tools. In the absence of detailed knowledge of the substances, first natural and then synthetic, it is just an operational classification. This is the case for conventional materials, because modern chemistry has introduced several classes of conventional materials such as metals and their alloys or ceramics, or materials derived from organic synthesis, such as polymers or liquid crystals. Recently, so-called advanced materials have been created: they no longer provide a passive response to external fluctuations such as heat or mechanical stress, but they provide an active response to one or more interactions, fluctuations or stimuli often of an electrical or optical nature. Known as functional or adaptive, they can transmit information and they are at the origin of more sophisticated multicomponent devices. The last stage of this evolution is that of nanomaterials, where the influence of surface functions will increase (the surface-to-volume or S/V ratio increases when size decreases). The nature and shape of the surface then become essential in these miniaturized devices.

1.3. Perspectives

The construction of an object requires some artistic creation when defining its shape whatever the type of material used. This is the case for a statue but also for a chair or a table to which a function must be associated, enforcing feet for the example cited. In the current context, this aesthetic component, which is always present, is called design. However, it is the transition from the aesthetic object to the technical object as an artifact that marks the boundary [SIM 58]. It is becoming more and more evident for highly technologically complex objects to be associated with ever more elaborate tools. This shape complexity will be extended by increasingly sophisticated functions. At the outset, this object has bulk properties that are described in physics (of the typical solid) as a thermodynamic system close to equilibrium. They will be exploited through energy transfers or exchanges to intelligently communicate with the environment. It is the nature of its separation surface with an appropriate shape and selectivity obtained by suitable treatments that will make it possible to define the materials and classify them as thermodynamic systems with a specific response. Prior to this, we will review the main achievements of phenomenological thermodynamics, the basis of this analysis.

2

Thermodynamics of Condensed Matter

A material, originally formed of specific chemical constituents, is used for a chosen purpose; it will become a specific object. As we saw in Chapter 1, we must define an appropriate shape and introduce a complementarity between form and matter, a paradigm of Western thought since Plato and Aristotle. This approach is reflected in the creation of an object with a specific use. It will depend on the intrinsic properties but also on the shapes and characteristics of its surface. The nature of the contacts and exchanges at the interface will define the field of application of a device, which is thus connected to the outside world. A proposal for classification based on the principles of thermodynamics thus appears as a general approach. Indeed, the conditions of use depend on the different constraints and flows exerted that define the behavior of the resulting material. It is therefore defined as a thermodynamic system subjected to exchanges of energy, and perhaps matter, within an environment that is considered to be stable.

We will introduce this in the first part of this chapter. To do this, it is necessary to review the essential principles of phenomenological thermodynamics at equilibrium and in particular, the concepts of energy and entropy. Then, in the framework of thermodynamics without equilibrium conditions, we will generalize these concepts and examine the response of the condensed matter, considered as a physical or chemical evolutionary system. As an example, the formation of a solid by kinetic processes of crystallogenesis and morphogenesis will be explained. We will discuss this in the final part of this chapter to make a connection with the selected field of application.

2.1. Definitions in thermodynamics

It is useful to review the general concepts and definitions used in phenomenological thermodynamics, which are indispensable when classifying a material. They are based on a set of axioms or principles, which we shall briefly review [MAR 95].

2.1.1. *Concept of a thermodynamic system*

The initial approach is to model the object of study by defining a system, the environment and their separation surface. The system, in its simplest form, is a so-called homogenous phase. This phase is normally composed of atoms or molecules, in short-range interaction with one another. This macroscopic system is then characterized by physical quantities called state functions.

A physical limit or boundary separates the given system from its environment. It can be deformable, subjected to a static or dynamic stress and behave like a selective filter. Finally, an observer records a response using various appropriate measurement techniques which, in turn, may influence the behavior of the system.

In a general context, there is a transfer of energy and/or matter, which depends on the relatively selective permeability of the boundaries for selected incoming and outgoing flows, respectively, counted positively or negatively. They are conditioned by the existence of a reservoir initially deemed very large and inexhaustible. An outgoing flow can affect the initial conditions by creating a feedback loop with an amplifier or buffer effect on the recorded response. This phenomenon leads to the distinction between constraints or flows that are permanent and those that fluctuate over time. It is then possible to obtain the corresponding responses to the either stationary or temporarily transient situations.

This simple case, historically proposed by Gibbs, is presented in Figure 2.1. It can be detailed with the presence of several reservoirs corresponding to various specific flows where the observer will acquire information by measuring the resulting properties or productions. The system, assumed to be homogeneous, may be in an isothermal or isobaric state, that is to say, at constant temperature or pressure or in an adiabatic state if it does not exchange heat with its environment. It may even be in a more complicated state, of a non-uniform heterogeneous nature, comprising several phases in the thermodynamic sense with many subsystems to be considered.

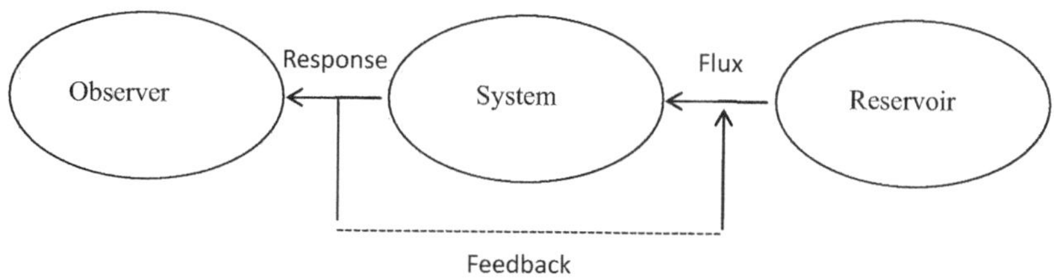

Figure 2.1. *Simple modeling of a thermodynamic system and its environment*

Then, the generalized thermodynamic approach leads to the definition of different classes of systems according to their interactions with the environment [BOC 68]:

– Isolated system: there is no exchange with the outside and the system tends toward thermodynamic or thermostatic equilibrium, no temporal evolution is observed.

– Closed system: only a flow of energy at the separation surface is present and this disturbs the initial equilibrium; the possible changes concerning the nature of each constituent can only arise from possible internal chemical reactions.

– Open system: there is both an exchange of energy and a transfer of matter, distinguishing the two flows, which as a result neglects the principle of mass–energy equivalence stated by Einstein with the generalization of the energy concept.

In this closed or open situation, a stationary state can be observed, which depends on the action of external driving forces and associated flows. These are often constant but can be time dependent; the first situation will be tackled first.

2.1.2. *Review of thermodynamic equilibrium states*

It is a system that obeys the first two principles of thermodynamics: laws of conservation of energy (U) and optimization of entropy (S) introduced by Clausius and characterizing the disorder of a system [BOC 68]. The latter can only increase and brings us to the general concept of an arrow of time and irreversible evolution.

In order to characterize such a system, thermodynamic potentials are introduced: these are extensive additive variables, which in an energetic representation are defined as the conjugates of the intensive parameters; temperature, pressure and chemical potential, respectively (T, P, μ).

For infinitesimal variations, we can write the following definition equations:

– T = dU/ dS, in an ideal reversible process in the absence of work, the associate heat element is dQ = T.dS, which corresponds to the increase in entropy dS that is a total differential;

– P = – dU/dV, corresponding to elementary work such that dW = –P.dV;

– μ = dU/dN, where N is the variable number of molecules for a chemical element present in the system, resulting from an addition or removal of material or during an internal chemical reaction.

Other external stresses or fields may act as intensive parameters, in particular electric (E) and magnetic (H) fields or even an electromagnetic wave as well as respective variations in the electrical charge or the associated induced magnetization (see Table 2.1).

Among the extensive state functions defining the system from a given set of energetic variables (free energy, free enthalpy or the grand potential function used subsequently), free enthalpy or Gibbs free energy (G) has a general form:

$$G = U + P.V - T.S + \Sigma\, \mu_i.N_i \qquad [2.1]$$

For a system with i distinct chemical components. The second principle then leads to the following expression of the variation in entropy outside a state of equilibrium:

$$TdS = dU + PdV - \Sigma\, \mu_i.N_i \geq 0 \qquad [2.2]$$

Reviewing the minima of these state functions and their total derivatives makes it possible to establish the conditions of thermodynamic stability and the existence of phase transitions in an equilibrium diagram elaborated according to the intensive parameters [BOC 68].

Finally, the fundamental physical quantity is energy, which can be presented in different forms by generalizing the concept on the basis of the historical equivalence between heat and work and the transformation efficiency of a thermal machine (the Carnot cycle). It is a chameleon physical variable, multifaceted but globally preserved by virtue of the first principle. Note that energy exists in six main forms: thermal, mechanical, electrical, radiative, chemical and nuclear.

As indicated at the beginning of this section, it always results from the product of an intensive physical quantity as variable and an extensive one. In Table 2.1, we have presented the main relations, in particular in the presence of an external field. Let us note that we have introduced the case of electromagnetic radiation at frequency ν for which the response of the system is the complex dielectric function (ε = n + ik with n refraction index and k absorption coefficient). The Poynting vector is a flow of energy expressed in units of time and surface, and the electromagnetic energy entering a volume element is the divergence of this vector.

Energy	Intensive variable	Extensive variable	Relation (of definition)
– Thermal:	Temperature T	Heat Q	$dQ \geq T.dS$
– Mechanical:	Pressure P	Volume V	$dW = -P.dV$
– Isotropic volume:	Stress ε	Elastic deformation σ	$dW = \varepsilon.\ d\sigma$
– Surface:	Surface tension Γ	Surface s	Surface energy $dG_s = \Gamma.\ Ds$
– Chemical:	Chemical potential μ	Chemical affinity A_c	Free enthalpy of
– Electrical (conductors):	Electrical potential Φ	Electric charge q	Reaction ΔG_r< or > 0 $dE = \Phi.\ dq$
– Magnetic (non-permanent magnetization):	Magnetic field H	Magnetization M	$dE_m = H.\ dM$
– Radiative:	Electromagnetic wave E(ν) and H(ν)	Complex dielectric function ε	Divergence of Poynting Vector: $\mathbf{S} = \mathbf{E} * \mathbf{H}$

Table 2.1. *Expressions of the different forms of energy in a system according to the main intensive parameters and external fields*

2.1.3. *Energy transformations and efficiency*

2.1.3.1. *Historical point: Watt steam engine*

After the work of several pioneers, the technological breakthrough is generally pinpointed to the end of the 18th Century with the invention of the Watt steam engine, making it possible to transform heat into mechanical energy. An example of a thermal machine is given in Figure 2.2, which shows how a double-acting device makes it possible, by means of various mechanical inventions, to induce a circular motion.

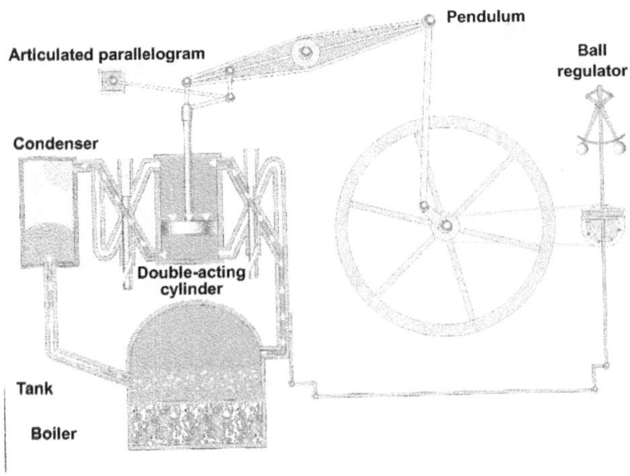

Figure 2.2. *Principle of the Watt double-expansion steam engine showing the different thermal and mechanical parts used to rotate a wheel*

This industrial revolution gave rise to thermodynamics with Carnot's memoir in 1824 entitled "*Réflexions sur la puissance motrice du feu*". This new science was developed during the 19th Century, in particular due to work by Clausius who established the concepts of energy and entropy.

2.1.3.2. *Conversions of different forms of energy*

The generalization of transformations or conversions from one form of energy to another took place following this precursor work by Carnot. In Figure 2.3, we review the main forms of energy, placing electricity at the center, a secondary energy that is not readily available in nature but essential for everyday use. Its introduction dates back to the end of the 19th Century with the discovery of electrochemical cells and then the birth of electric motors [UPT 80]. It can be obtained from one of the primary energies, available through multiple access routes. In this figure, we have indicated, by means of example, the most important types of transformations. They are represented by a simple arrow for natural or anthropomorphic conversions from fossil or sustainable resources. These are essentially:

– solar radiation giving rise to photosynthesis of plants, the basis of life on Earth, or the absorption by a black object transforming radiative energy into heat;

– transformation of matter by thermochemistry, combustion of fossil fuels, coal, oil or gas, or resulting from nuclear fission or fusion;

– thermomechanical machines transforming heat into work that have existed since the invention of the steam engine (see Figure 2.2).

We have also represented the processes leading to the creation of electricity using double arrows. Note the techniques and processes based on thermoelectric effects (heat gradient), turbines and wind turbines associated with mechanical forces (waterfalls, tides and winds), photoelectric modules (solar radiation) and electrochemical transformations (batteries, fuel cells). These devices, which we will discuss in the following chapters, all use isolated or assembled materials.

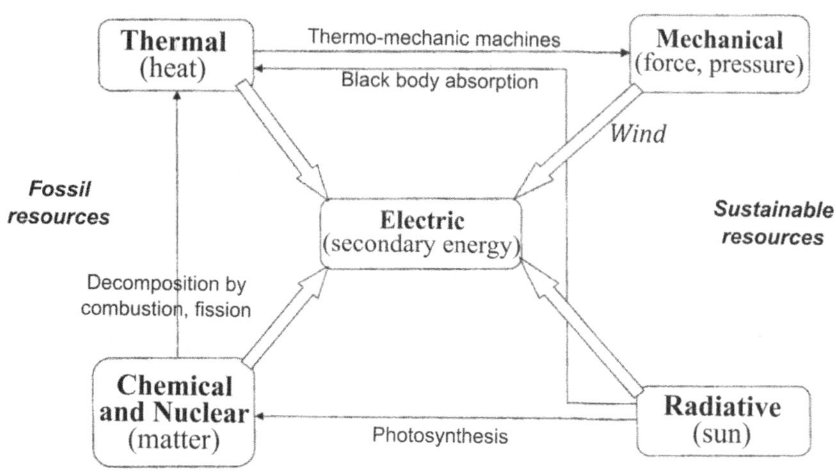

Figure 2.3. *Examples of energy conversions between the main primary energy sources and leading to the generation of electricity*

These transformations, whether direct or not, are carried out at a certain yield, because there is always dissipation of heat, the most degraded form of energy. Their efficiency is translated by a conversion factor η between 0 and 1 corresponding to the energy comparison of the initial and final state:

$$\eta = E/E_0 \quad [2.3]$$

where E_0 is the energy available at the input and E the energy obtained in a different form and recovered after transformation. Thermal machines with two reference temperatures are historical examples of an energy conversion.

Thus, the maximum proportion of heat that can be converted to work depends exclusively on the temperature of the cold source at the outlet (T_2) and that of the hot source at the inlet (T_1):

$$\eta = 1 - T_2/T_1 \quad [2.4]$$

This ideal conversion factor can be calculated and compared to the experiment for all possible transformations from a couple of intensive variables. It is called a thermodynamic limit because it is optimal for an ideally reversible transformation. In practice, this is not the case, the actual yield is not known in advance and the concept of delivered power, energy per unit of time, becomes a key concept. It is the rate of energy release, homogeneous to an energy flow, which can be relatively quick. Controlling dissipation mechanisms in the form of heat from other forms of energy thus becomes a crucial point:

– mechanical, involving the friction or shock of moving parts, it is the friction mechanism studied in tribology that can give rise to other transformations (triboluminescence, triboelectricity);

– chemical (or nuclear), heat released by an exothermic reaction (or endothermic, which requires an energy supply to initiate it);

– electrical by the Joule effect, the power transformed into heat is $W = R.I^2$, where R is the electrical resistance of the material crossed by a current of intensity I;

– radiant, the law of conservation of energy leads to equality: $A + T + R = 1$, where A is the absorption and T and R the transmission and reflection of the electromagnetic wave at the interfaces with the surrounding environment.

Absorption can only lead to a conversion of energy, for example, in photosynthesis phenomena or photovoltaics also associated with the production of heat. This degradation is optimal in the case of the black body ideally defined for A = 1. In this case, there is absorption and corresponding emission, a function of the temperature reached (Stefan–Boltzmann law).

These observations led to the technical concept of available or usable energy introduced in exergy analysis. This energetic approach makes it possible to deal with the practical problems of energy management [BOR 05], which are based on the exergy, the practically useful fraction that is not conserved. The unused part (called anergy) is related to the production of entropy in an evolutive system: this is introduced in the following section.

2.1.4. *Systems without thermodynamic equilibrium*

For an extensive variable or a thermodynamic parameter E such as an energy, we must distinguish the source, the inherent creation of the system, from the flows, referring to the exchanges with the outside world [MAR 95]. A global balance equation is written from the temporal variation expressed as a total differential:

$$dE/dt = \int_v \sigma(E).dv + \int_\varepsilon J(E).d\varepsilon \qquad [2.5]$$

where σ (E) is the source and the integral is defined for the volume v of a continuous medium such as a production of energy P(E), and J (E) is the flux density integrated over a separation surface or boundary ε.

The research into stationary states (dE/dt = 0) is essential. For an isolated system, this term is zero in accordance with the first principle of conservation. In the case of a closed or open system with increasing exchanges, a stationary or oscillating state can be obtained under conditions of stability. This non-equilibrium thermodynamics was largely developed during the 20th Century, especially by Prigogine and his school on the basis of the presence of a local equilibrium [PRI 68]. Two different situations can be encountered as one moves away from the equilibrium conditions.

2.1.4.1. *Irreversible linear processes*

Irreversibility is a consequence of the second principle in an isolated system; in general, it is necessary to specify the entropy balance concept for closed or open systems. From the general equilibrium equation (equation [2.5]), the variation in entropy during a time interval dt is split into two terms:

$$dS = d_iS + d_eS \quad [2.6]$$

where diS is the internal production of entropy, positive or zero and deS is the flux of entropy due to exchanges with the outside.

In the stationary state:

$$dS = 0 \text{ and } -d_eS = d_iS \geq 0 \quad [2.7]$$

Several types of situations exist, which are functions of the nature and intensity of the flows imposed on the system and of the internal production generated in particular by the presence or absence of chemical reactions. It is the criterion of evolution for which the variation over time of the entropy expressed per unit volume in a continuous medium is written by integrating over the volume of the system considered [GLA 71]:

$$dS/dt = \int_V \sigma s\,(r,t)dV - \int_V di_V.J_s\,dV \quad [2.8]$$

This balance equation corresponds to the existence of an entropic flux Js expressed by a divergence operator and an internal source of entropy σ_s per unit volume and time. Note that during an irreversible process, the external flow can counterbalance the internal production of entropy. It may be linked to a supply of electrical, radiant, thermal or mass energy.

This stationary state ($dS/dt = 0$) is, therefore, characterized by the production of entropy in a situation close to equilibrium and is thus represented in the linear approximation when the effects are proportional to the causes:

$$P\,(S) = d_i\underline{S}/dt = \Sigma\,\mathbf{J_k.F_k} > 0 \quad [2.9]$$

where **J** and **F** are the flux vectors and the generalized forces used for the elements k, respectively. Linear laws are expressed by Onsager relations [BOC 68], which we will discuss in section 2.1.4.2. In this context, Glansdorff and Prigogine established a minimal entropy production theorem for diffusive systems [GLA 71]. However, the domain of validity in a linear regime is limited, when the rate of the exchanges increases the resulting steady-state regime can become unstable ($d\underline{S}/dt \# 0$).

2.1.4.2. *Nonlinear processes*

When a system, subjected to increasingly imposed flows, is further away from equilibrium, a nonlinear response is observed. This is seen for a critical or threshold value with the appearance of new so-called dissipative structures that are characterized by instabilities creative of order. Indeed, assessing the stability criteria for a local variation of entropy (expressed using equation 8 for an elementary volume) shows a change in response, which can be associated with a discontinuity in the slope of the internal production of entropy [GLA 71]. A more general theory of thermodynamics beyond the linear domain has been established from the inherent fluctuations in a system experiencing increasing fluxes. When one moves away from equilibrium, fluctuations in energy or matter increase and they can move from a regressive to a progressive situation, creating a form of instability. A new phenomenon appears for a critical value mathematically corresponding to a singular bifurcation point (see Box 2.1 and Figure 2.5) with the creation of a new non-uniform temporal structure or several possible stable states. The fields of exploration of these phenomena mainly cover fluid hydrodynamics but also electrical circuits (coupled oscillators), optics (lasers) and chemical kinetics in a condensed medium [VID 88]. As for fluid dynamics, the Rayleigh–Bénard instability leads to the formation of convective rolls observed under a strong thermal gradient or the experience of the circular Couette flow of a fluid layer subjected to intense shear: they can develop up to a dynamic state of motion that becomes turbulent [BER 84]. In these dissipative structures, a new order can appear and the dynamics of the system are then governed by an extremum principle of maximum entropy production. As proposed by different authors, an evolutionary system selects processes that minimize their local potential function and optimize the entropy variation to obtain maximum outputs [DEW 05]. This maximization of the temporal derivative of entropy (equation [2.8]) is developed in the context of non-equilibrium statistical physics (see Chapter 5).

2.1.4.3. *Deterministic chaos*

We must complete this presentation with the possible appearance of deterministic chaos in these dissipative systems. It is a strictly dynamic aspect showing how an unpredictable situation can arise in a relatively simple system with a limited number of degrees of freedom; this is due to the inherent uncertainty of knowledge about the initial conditions of the system.

These recent achievements concerning chaos theory will only be briefly mentioned. The appearance of a chaotic situation rests on three possible routes: the passage of a signal with periodic oscillations toward frequency doubling, the appearance of quasi-periodicities or an aperiodic signal, or even via a so-called intermittence mechanism. They are developed by a Fourier analysis characterizing a limit cycle for sustained oscillations and theoretically by Lyapunov functions that define a dynamic state [BER 84]. A cascade bifurcation diagram with a sequence of transient states up to a chaotic state is proposed to account for this. The final regime obtained is linked to the appearance of an attraction pool linked to the existence of different types of attractors. In particular, the mathematical concept of a strange attractor is associated with deterministic chaos when the trajectories of the dynamic system are confined to a particular area of the phase space (space of positions and particle impulses, see Chapter 5). This concept is derived from Poincaré's century-old work on the theory of these dynamic systems applied to celestial mechanics. The evolution of these systems is not known in a univocal way since the trajectories observed are sensitive to the initial conditions or to the subsequent perturbations. Thus, different scenarios exist to reach a deterministic chaotic state defined using the Lyapunov stability equilibrium, equivalent to Gibbs expressions near to equilibrium [SCH 05].

Finally, it should be noted that the first physical model proposed was that of Lorenz in 1963 for meteorological forecasts. To account for the turbulent behavior of the Earth's atmosphere, this author had proposed a set of three nonlinear differential equations that allowed for a simple numerical simulation of such a dynamic situation [BER 84].

2.2. Examples of hardware systems

From these general considerations, we shall examine the behavior of various systems consisting of condensed phases in the absence of chemical reactions, and then consider their particular influence. Note that a chemical reaction can be defined by the destruction and creation of chemical bonds associated with an exo- or endothermic transformation energy (see Table 2.1).

2.2.1. *Responses close to equilibrium*

If the intensive parameters vary, the system is no longer in a timeless state of equilibrium; there are associated energy fluxes, which are a function of the thermodynamic forces present. When these are weak, we can consider, in a first approximation, a stationary relation where the observed fluxes are linear responses:

$$\mathbf{J}_i = \Sigma\, \mathbf{L}_{ij}.\, \mathbf{F}_j \qquad [2.10]$$

Onsager has shown that the relation between these vectorial quantities is a symmetrical matrix of phenomenological coefficients $\mathbf{L}_{ij}$ with symmetry conditions obeying the Curie principle (the symmetry of the causes must be greater than that of the effects). This first-order approximation expresses a large number of empirical laws of great practical importance for the study of the properties of condensed materials [BOC 68]. When there is a single force present, there are different macroscopic laws:

– Ohm's law, in the presence of an electric potential gradient, is the electrical conductivity of a medium that can carry electrical charges;

– Fourier's law, in the presence of a thermal gradient and expressing the thermal conductivity, i.e. the transport of heat in a solid;

– Fick's law, in the presence of a concentration gradient of a chemical component and the associated mass transport by a diffusion mechanism.

2.2.1.1. *Isothermal electrical transport*

For a solid in an isothermal regime, with an electronic conduction and therefore a flux of charges, electrical resistivity is a typical characterization technique. It gives an account of the proportionality between the potential difference and current intensity, thereby defining an electrical resistivity. As shown in Figure 2.4(a), each sample is thus a closed system exchanging electrical energy with the exterior. The importance of the geometry, shape and section of the samples necessary to carry out a significant physical measurement should be noted. In the presence of an electric field and a magnetic field, two galvanomagnetic phenomena are observed due to the presence of Lorentz forces [BOC 68]. On the one hand, the magnetoresistance, the relative increase in the electrical resistivity normally giving a positive effect (by increasing the dissipated heat and the associated entropy), which is a function of the orientation of the applied magnetic field

(Figure 2.4(b)); on the other hand, the Hall effect, the appearance of a potential difference in the direction perpendicular to that of the current flow in the presence of an orthogonal magnetic field (Figure 2.4(c)).

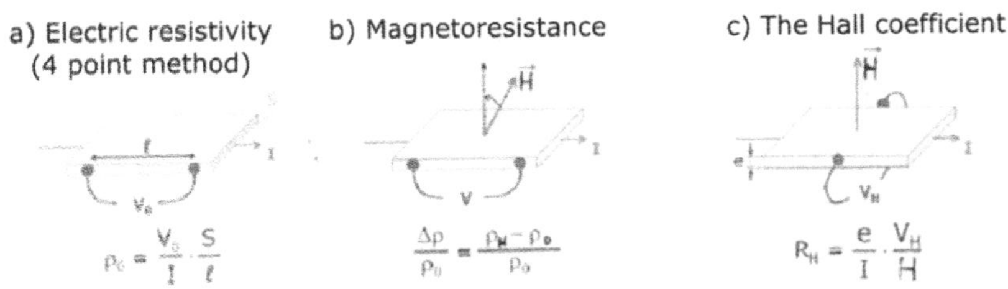

Figure 2.4. *Experimental measurements of the isothermal electric transport on a parallelepipedic sample whose dimensions S and l are indicated on the figures: (a) electric resistivity ρ_0 for a current with intensity I and a difference in voltage V_0; (b) magnetoresistance $\Delta\rho/\rho_0$, relative variation in electrical resistivity in the presence of a magnetic field* ***H****; (c) Hall coefficient R_H linked to the appearance of a transverse component of voltage V_H in the presence of orthogonal fields* ***E*** *and* ***H***

In the case of two or more acting forces with coupled flows, several interrelated effects exist. Onsager's reciprocal relations make it possible to account for many phenomena from the expression of kinetic coefficients associated with the combination of different fluxes [BOC 68]. Let us mention other major effects in relation to energy conversions and materials.

2.2.1.2. *Thermoelectric effects [PRI 68]*

In the presence of a thermal gradient in an electrical conductor, three phenomena have been explained:

– the Thompson effect, additional heat dissipation in a conductor subjected to a temperature gradient, contributing to the usual Joule effect;

– the Peltier coefficient, reflecting the evolution of heat accompanying the flow of electric current through an isothermal junction of two materials;

– finally, the Seebeck effect, production of an electromotive force under zero current in the presence of two junctions at different temperatures; this device is called a thermocouple.

These three effects are interrelated by Kelvin's relations.

2.2.1.3. *Electrokinetic effects [PRI 68]*

In a fluid system separated by a porous wall, the temperature and concentration are uniform but the pressure and electrical potentials could be different. The interdependence of these processes leads to the appearance of a flow potential, an effect of electroosmosis flow of matter under an electric current, and the existence of an electroosmotic pressure with a flow current. Saxen's relations link the osmotic effect to the flow effect.

2.2.2. *Responses far from equilibrium*

Many situations exist and we will focus only on interesting cases. They depend on the way the matter is transported, here by diffusion only, and in the presence of coupled chemical reactions [GLA 71]. We are simply interested in the competition between nonlinear chemical reactions due to an autocatalytic effect and molecular diffusion in so-called reaction–diffusion systems. A reminder of the theoretical bases is given in Box 2.1.

It is an isothermal reaction system whose evolution is governed by chemical transformations where the transport of the associated species is ensured by molecular diffusion (Fick's law). A phenomenology of nonlinear chemistry has been gradually established, which accounts for these phenomena [DEW 09a]. It is a competition between a scalar process, the reaction rate constant and a vectorial process, the diffusion related to the concentration gradient. Mathematically, the dynamics of the system depend on each chemical species i or X_i ($\mathbf{r}$,t) and governed by partial differential equations in the following form:

$$\delta X_i / \delta t = f_i[\{X_i\}] + D_i \, . \, \mathbf{\Delta} X_i \qquad [2.11]$$

where f_i represents the rate of chemical transformation of $\{X_i\}$, D_i is the diffusion coefficient and $\mathbf{\Delta}$ is the associated Laplacian operator.

Box 2.1. *Reaction–diffusion systems*

These equations are complemented by boundary conditions of the system that depend on external factors. In order to obtain reaction–diffusion structures, concentration non-uniformities must exist over time scales greater than the characteristic diffusion times. To do this, we must reach a critical threshold characterized by a bifurcation point where there is a two-branched separation (see Figure 2.5), which induces an oscillating phenomenon or the possibility of bi- or multistable states.

There are different types of bifurcations, such as the Hopf bifurcation, where temporal structures can occur [BER 84]. There are other types of bifurcations and an existence diagram for thermodynamic states far from equilibrium can be established according to the stresses and flows present. Thus, the transition from an oscillating system to a bistable system can occur by modifying a feedback loop as a function of the imposed flows [DEW 09].

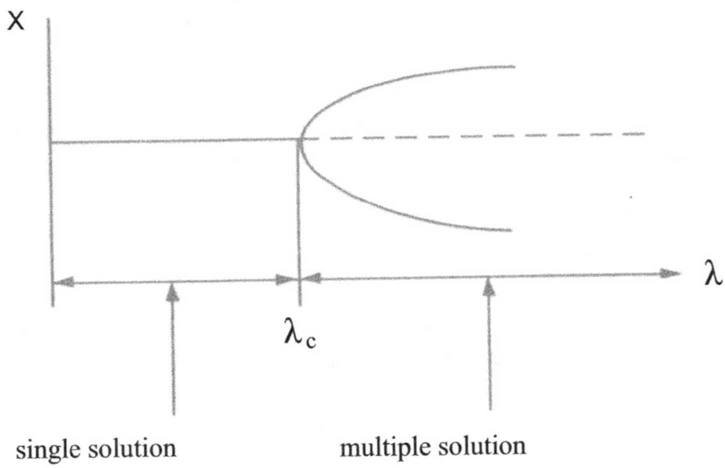

Figure 2.5. *Bifurcation point corresponding to the diagram presented in the simple case of a function $X = d\lambda/dt = r\lambda - \lambda^3$ with $r \geq 0$ and the critical threshold λ_c = + or – ($r^{1/2}$)*

This type of self-organized system is part of a set of models derived from the general theory of Langevin where two processes are antagonistic: the fluctuating Brownian motion and the particle's force of flow. In this context, a different approach has also been established for critical binary systems with phase demixing phenomena close to equilibrium [DES 09]. Phase segregation and spinodal decomposition kinetics cause a break in symmetry with the appearance of spatial patterns similar to those obtained with the Turing model [TUR 52] as described in the text, even though the thermodynamic situations are different. These are Cahn–Hilliard models, based on the minimization of the free energy of a system subjected to a diffusive regime with the competition between the mixing entropy and the minimization of the enthalpy related to the favorable sorting of identical molecules.

Experimentally in an open homogenous reactor, two main instabilities occur. With regard to the experimental conditions, when only diffusion processes are present, they depend on the level of the flows of matter which characterizes a residence time of the reactive species; two types of results are observed:

– Periodic chemical oscillations or chemical waves with the existence of a stable limit cycle. This is the typical example of coupled redox processes in a Belousov–Zhabotinsky reaction and other similar examples [DEW 09a]. In this case, the periodic oscillations shown in Figure 2.6 are associated with a color change in the solution where citric acid is oxidized by potassium bromate in the presence of cerium ions as a catalyst. In the example given, the Fourier analysis reveals harmonic frequencies for these concentration and flux conditions. This is an initial step toward the desynchronization of oscillations and the subsequent appearance of a chaotic state.

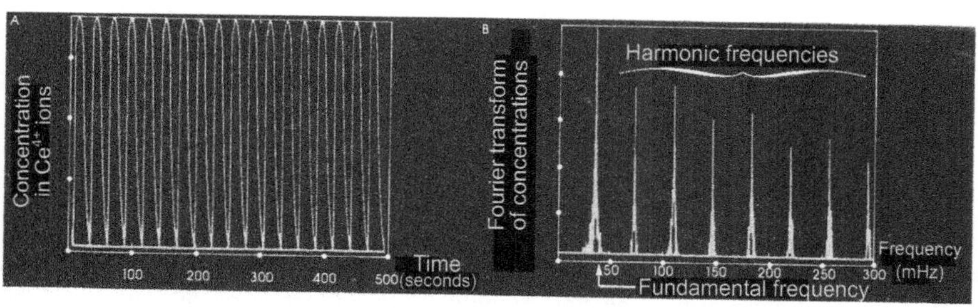

Figure 2.6. *Example of periodic oscillations in the Belousov–Zhabotinski reaction obtained by recording the time variation of the concentration of the cerium ion (A); the signal obtained is no longer as periodic and the analysis of the Fourier spectrum (B) shows the appearance of harmonic frequencies in addition to the fundamental one (adapted from [VID 88])*

This oscillatory behavior, of a frontal dynamic with propagating waves, was modeled from equation [2.11]. Numerical simulations show that a limited number of coupled reactions and degrees of freedom can induce these phenomena. This is the case, for example, with the "irreversible

Bruxellator", a numerical model where the reaction comprises only four nonlinear reactions coupled together [VID 88].

– Spatial bistabilities corresponding to another type of bifurcation and the appearance of stationary structures laid down by Turing [TUR 52]. Under certain conditions of autocatalysis, different domains with a stable structure of concentration fronts are observed. They correspond to a bistable state with two branches, one thermodynamic and the other dissipative or flux, separated by an unstable transition zone with a hysteresis loop. This second branch appears for short residence times in the reactor. A schematic illustration of the phenomenon in a chemical model with this behavior is shown in Figure 2.7.

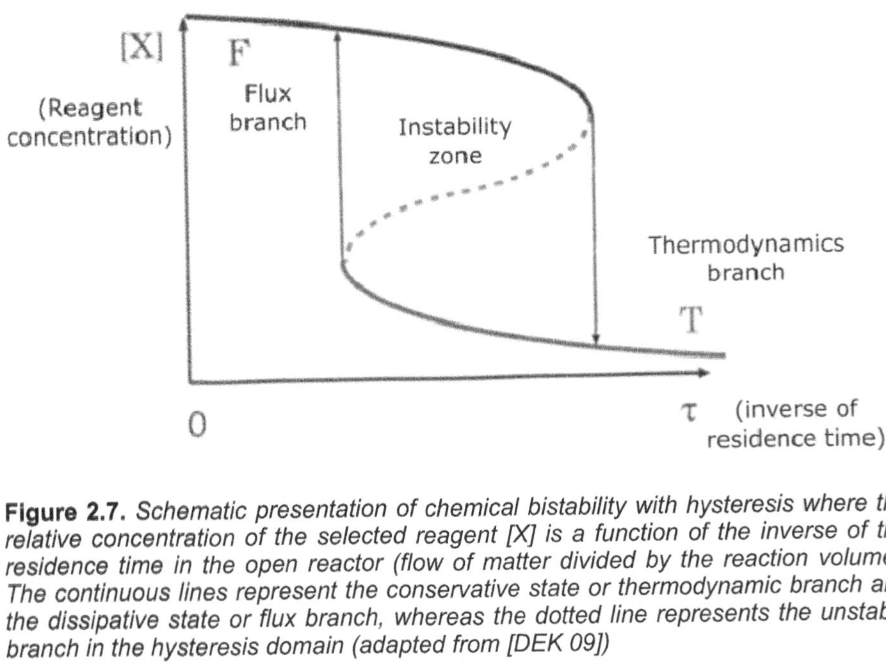

Figure 2.7. *Schematic presentation of chemical bistability with hysteresis where the relative concentration of the selected reagent [X] is a function of the inverse of the residence time in the open reactor (flow of matter divided by the reaction volume). The continuous lines represent the conservative state or thermodynamic branch and the dissipative state or flux branch, whereas the dotted line represents the unstable branch in the hysteresis domain (adapted from [DEK 09])*

Thus, in these systems, which are kept homogeneous and open, subjected to increasing circulations of matter, a self-organized regime appears with the existence of distinct dynamic oscillating or stationary situations. They are described in a relatively complicated existence diagram including possible combinations and transitions [DEK 09].

2.2.3. *Role of chemical reactors*

In a thermodynamic system, energy involved in chemical reactions plays a particular role. Note that in the linear approximation, the chemical affinity (A) is the extensive variable acting as a generalized force for the reactions to proceed with entropy production in the system considered. Also note that this approximation is no longer valid for flux values above a critical threshold [DEW 09a].

We will describe the different situations encountered for an experimental chemical reactor considered as a thermodynamic system, defined with a real enclosure and subjected to constant contributions of energy and/or matter. For this purpose, on the basis of Figure 2.1, we represent a reactor, like those designed in chemical engineering, to analyze its operation. Figure 2.8 shows a reactor with a reservoir for inputs of matter and a thermostat that can regulate the temperature due to incoming or outgoing heat flows. Only the exchange of heat, a degraded form of energy, is taken into account but several situations are possible.

2.2.3.1. *Isolated system*

No exchange of energy or matter with the exterior occurs – this is the principle of an adiabatic calorimeter. By measuring the associated temperature changes, it is possible to determine the enthalpies of reactions or those associated with conventional phase transitions. It also allows for the measurement of the calorific capacity of a solid as a function of temperature, a physical quantity that gives information directly on the internal entropy variation of the system.

2.2.3.2. *Closed system*

There is no supply of matter from a reservoir; the reactants are initially added and the reactor then operates discontinuously. However, it is stirred and heated to ensure homogeneity of the reaction, which in general evolves regularly to reach the state of final equilibrium corresponding to its maximum entropy.

A thermostat then makes it possible to control the temperature due to exchanges of heat associated with the enthalpies of reaction. An isothermal situation is necessary in the case of kinetics, such as the measurement of a constant for reaction kinetics or the determination of a chemical equilibrium. If the reactor is in free running, in the absence of a thermostat, the exothermic chemical reactions will heat up and produce heat as a primary source of energy, such as a wood stove or a coal-fired power plant.

NOTE.– A heat exchanger with two independent flows of matter running in the opposite direction and in a stationary state shall be considered as a closed system.

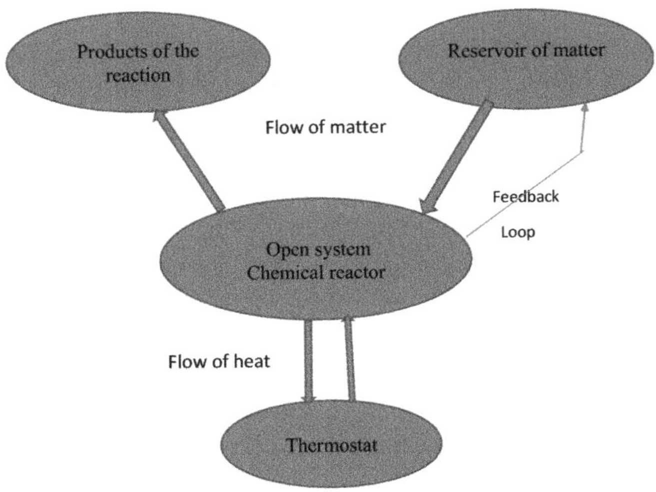

Figure 2.8. *Diagram of a chemical reactor considered as an open system able to exchange matter and heat with its environment*

2.2.3.3. *Open system*

A constant amount of matter flows continuously from the reservoir and the reaction products are recovered at the outlet (Figure 2.8). Two regimes exist depending on the input and the selected thermostat temperature, which controls the reaction rate (Arrhenius law). In a linear regime, the matter and heat flows make it possible to obtain a stable stationary state, which is often desired in continuously powered industrial reactors, for example in cement plants. Another significant case is the burning of fossil resources, which provides heat, which is then recovered to be transformed into more noble, mechanical or electrical energy by thermal machines (see Chapter 4).

A more unique situation is that of a nonlinear dissipative regime when the mass flows supplying the system are increasingly large and exceed a critical threshold. To control this situation, particular reactors need to be used. They must be continuously stirred or filled with a porous gel to ensure a single diffusion process and they then have at least one externally controlled interface [DEW 09a]. Coupled chemical reactions can lead to a bistable spatial state (Figure 2.7) or alternatively produce specific chemical species resulting in chemical oscillations (Figure 2.6).

The existence of stable Turing structures was demonstrated almost 40 years later by strictly controlling the operation of a self-sustaining isothermal reactor. An example is given in Figure 2.9 with a comparison between experimental observations and the result of numerical simulations [BOI 09b]. This example is significant because it allows us to glimpse the topological and even morphological analogies between the inert and the living world (a comparison of patterns observed on the coats of leopards and zebras). It suggests common mechanisms governing the rules of geometrical construction. Thus, chemical reactions can have creative rhythms and forms, inducing phenomena of spatial self-organization that are present in biology, involving several interacting chemical reactions. This comparison with living systems has been analyzed on the basis of the analogies of stable forms, which induce the possibility of a common origin [BOU 80]. This approach is in line with the seminal work cited in Chapter 1.

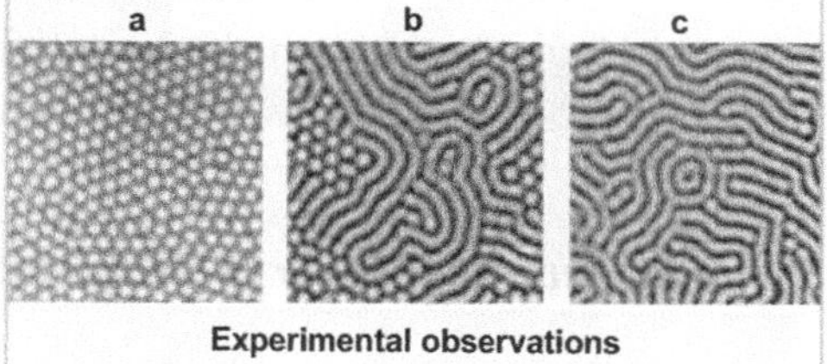

Figure 2.9. *Example of sequences of Turing structures obtained from the reaction of chlorine-iodine-malonic acid [BOI 09b]. In experimental observations (a–c), the concentration of malonic acid increases from left to right and numerical simulations of a kinetic model account for this evolution, ranging from the formation of spots to that of bands*

NOTE.– From an experimental point of view we noted that, in various cases of open reactors, a flow varying in time was not cited. Indeed, if the flow of energy or matter emanating from a reservoir is not stable, equally transitory and more complex situations may arise. This is particularly the case in living systems, which will be discussed in Chapter 9.

2.3. Material development and characterization

We shall detail the conditions required for the formation of a condensed phase, defined as a homogeneous atomic or molecular entity. For this, we are interested in the formation of an ordered structure of matter and then with real crystals, that is to say, containing network defects, or even, losing elements of crystallographic symmetry. This approach can lead us to phases with an intermediate state between order and disorder, which is particularly the case for molecular constituents arising from organic chemistry (e.g. polymers, liquid and plastic crystals, etc.). Finally, when the size scales increase, the study of the growth modes of inert forms can be approached by also considering chemical morphogenesis.

The existence of these solid phases will be complemented by a multiscale or hierarchical description. They are characterized by different observation and analysis methods such as optical and electronic microscopies. As we pointed out in the first chapter, the following dimensional scales can be distinguished:

– atomic scale, which defines the geometrical arrangements associated with the types of chemical bonds present between neighboring atoms;

– mesoscopic or textural scale, which is based on the formation of colloidal microcrystals or aggregates, and their spatial arrangement;

– macroscopic or morphological scale, which characterizes the shape and the possible symmetry of larger objects, and which can be visually observed.

2.3.1. *Situation close to equilibrium: crystallogenesis*

Crystal growth is based on surface thermodynamic phenomena. The overall nucleation-growth process involves the formation of a homogeneous or heterogeneous nucleus followed by different growth modes [GIL 63]. The simplest form of formation is homogeneous nucleation, the birth of a stable germ for a critical size reached when its Gibbs energy (equation [2.1]) is greater than the surface energy present, which is equal to the product of the surface created by surface tension (see definition Table 2.1). The properties of crystal symmetry are found in the outer shape of the crystal since the growth mechanisms depend on the surface energies of the different crystalline planes. The construction of a minimal energy surface can indeed lead to a polyhedron, which is a function of the elements of symmetry of the crystal lattice. An alternative pathway is heterogeneous phase nucleation, where the role of an external surface is to lower the free enthalpy of formation and thus the critical size necessary to obtain a stable germ. This mechanism is associated with the phenomenon of epitaxy, a phenomenon of oriented growth, often associated with specific chemical reactions in the presence of catalysts.

As a general rule, controlling the crystallization parameters makes it possible to obtain a crystallized solid whose kinetics of formation from a gas or liquid phase depends on the gradients of the intensive thermodynamic parameters already introduced. The transport of energy and matter in an open reactor close to equilibrium is governed by Onsager's linear relations. An example is given in Figure 2.10, showing an industrial technique for manufacturing monocrystalline silicon wafers for the electronics industry. The graphite crucible containing the liquid silicon acts as a reservoir for the growth of the single crystal. From a general point of view, there are three types of processes: vapor transport, the method of supersaturation of a solution, and heterogeneous nucleation from a monocrystalline germ: the latter is the case we are discussing.

However, it should be noted that an increase in temperature or pressure gradients can result in the appearance of a multitude of simultaneous nucleations and the juxtaposition of crystals of different orientations forming a polycrystalline assembly. The widespread technique of vapor deposition is a generalization of transport under low temperature gradient associated with chemical reactions. It leads to these multinucleation situations during the formation of a thin layer of an element or a solid solution.

The key point to consider is the presence of allotropic varieties for certain elements. These are different crystal arrangements called polymorphs with several possible minima of the free enthalpy (or Gibbs) state function. Being very close, they create stable or metastable states in the case of a primary or secondary energy minimum. A particular example is that of carbon, which may give rise to several crystalline phases: massive, diamond, graphite and carbyne phases according to the coordination number of the carbon element, or molecular tri-coordinated forms: fullerene, nanotube or graphene (see Chapter 6). Graphite is the stable phase under normal conditions of the intensive parameters T and P and diamond from a thermodynamic point of view is called a metastable phase. Thus, according to the experimental constraints, various crystalline phases as well as different textures on the nanometric scale are obtained [GOG 00]. An example of polyhedral crystals is shown in Figure 2.11, which recalls the earlier mineralogical descriptions on a macroscopic scale (see Figure 1.4).

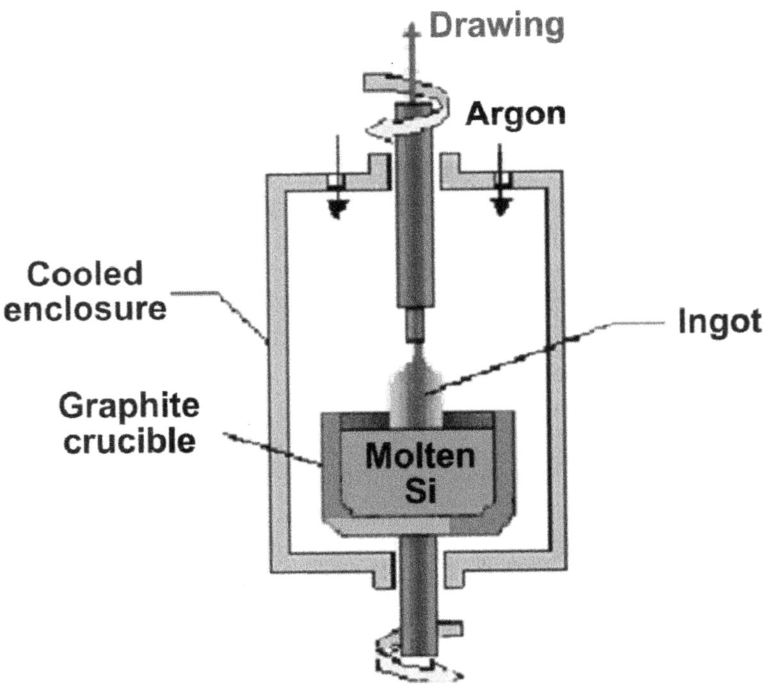

Figure 2.10. *Scheme for manufacturing a silicon ingot in a graphite crucible heated above the melting point of the element by the Czochralski method; it shows the priming and growth by drawing a crystal from a germ*

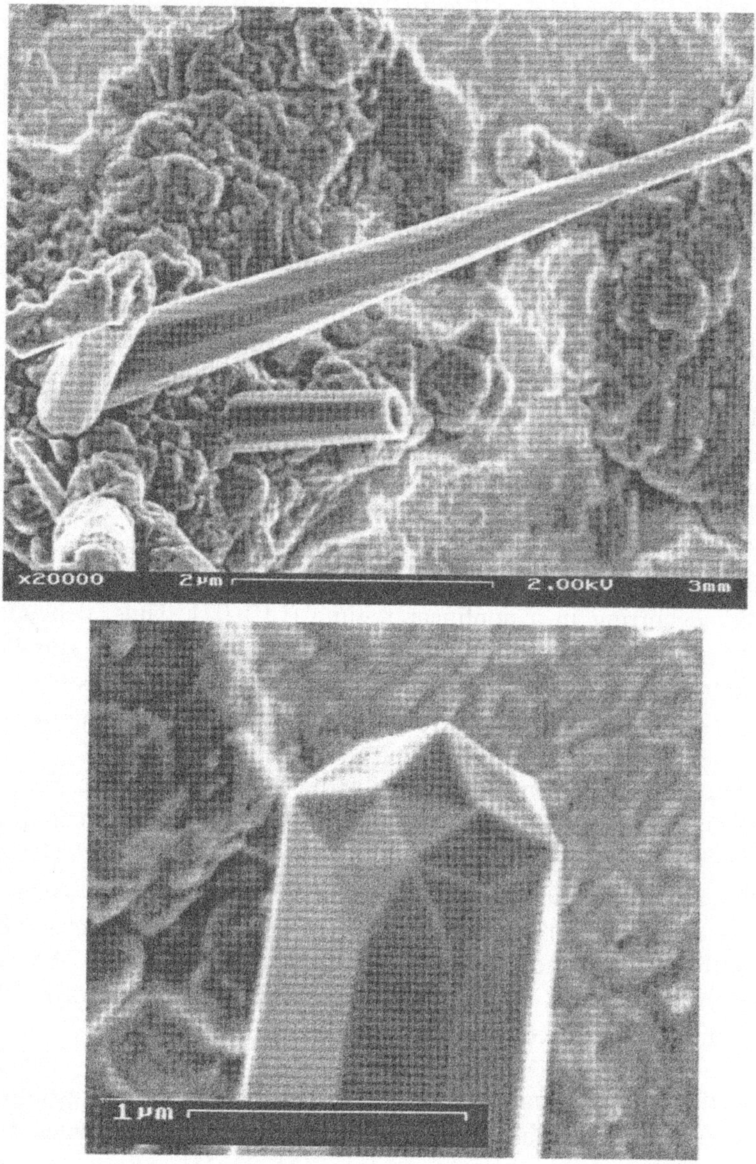

Figure 2.11. *Examples of microstructures or textures made using an oxygen-acetylene torch and observed by scanning electron microscopy and photographed at different scales showing the formation of a graphitic polyhedron (private communication and thesis of Hanako Okuno, Université catholique de Louvain, Belgium 2006)*

2.3.2. *Situation far from equilibrium: morphogenesis*

When the phenomenon of growth at the interfaces increases and becomes nonlinear, different morphologies appear, which have to be characterized at several scales. This surface increase mechanism involves strong or weak chemical bonds; the latter, which are more labile, lead to an optimization of the textural forms. Numerous systems have been studied as well as attempts to classify growth types, mainly from mass transport that may include convection in addition to diffusion. They have been proposed taking into account the actual interaction with the convection mechanism at the interface [PEL 00]. Two major approaches have been recognized; on the one hand, those associated with non-classical crystallization processes [PEL 00], as we have already discussed (see Figure 1.8), and on the other hand, those due to interfacial reaction instabilities generating different morphologies in particular dendritic morphologies [COL 08].

The formation of ice crystals is a key example (see Figure 2.12). It has been shown that the large increase in observed forms is due to crystallization instability occurring in a nonlinear regime [LIB 07]. Thus, depending on these two constraints, temperature and moisture content (water supply), it is necessary to distinguish between the domain formation of plates from that of dendrites, needles or columns corresponding to different growth modes.

In the case of a confined geometry and a fluid initial phase, the viscous digitization phenomena, during crystallization or electrodeposition in the presence of an electric field, give rise to more complex shapes. These more hierarchical textures are analyzed using diffusion limited aggregation processes and the formation of branched morphologies that may present a spatial self-similarity and a so-called fractal structure [PEL 00]. An example is given in Figure 2.13 which shows the electrochemical growth of metal particles in a confined geometry.

Thus, these forms become increasingly complex, no longer resulting from a simple flattening of plane crystalline surfaces as in crystallogenesis. The fractal geometry proposed by Mandelbrot [MAN 74] makes it possible to describe complex and irregular natural forms that do not have a traditional geometry. This mathematical modeling can be applied to various fields of research, ranging from physics and biology to economics, and even describe deterministic chaos as we shall see later.

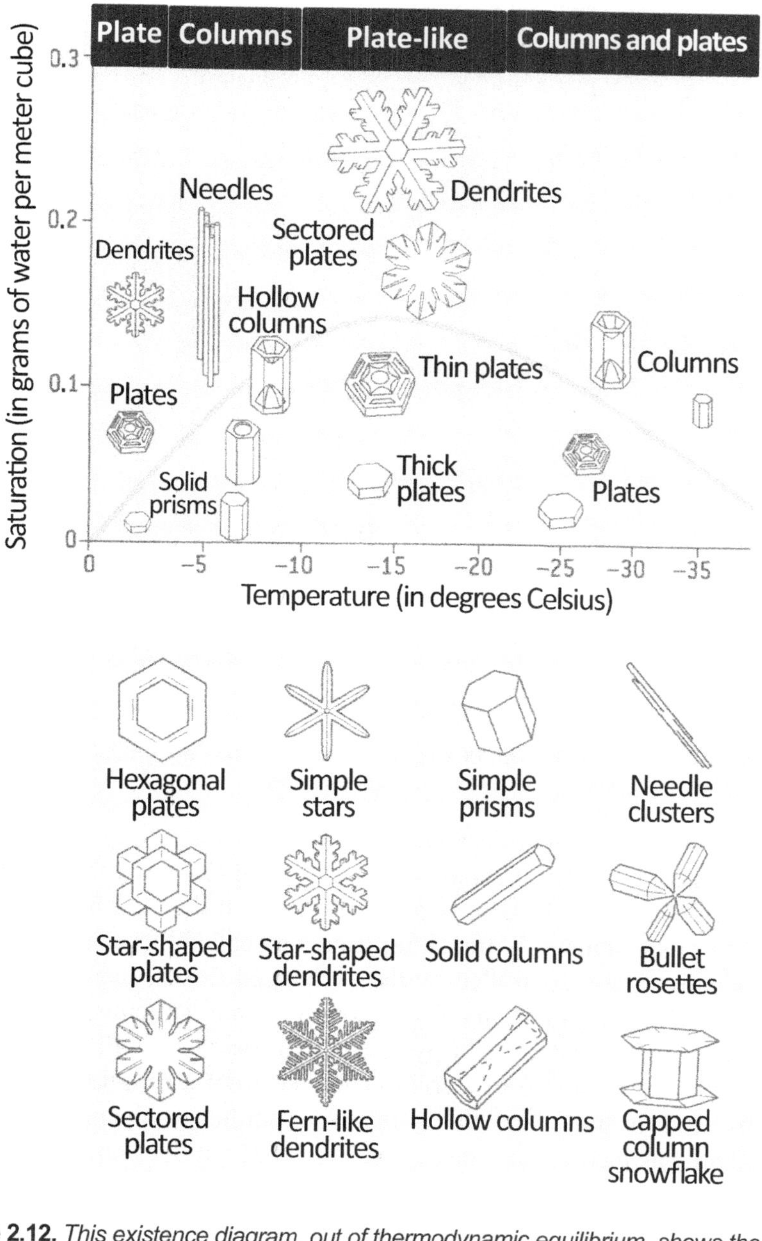

Figure 2.12. *This existence diagram, out of thermodynamic equilibrium, shows the great variety of forms observed in snow crystals as a function of supersaturation when moisture exceeds 100% and the ice formation temperature (according to [LIB 07])*

Figure 2.13. *Branched growth of a copper deposit obtained by electroplating in a two-dimensional cell of 50 µm thickness (according to [ARG 88])*

With regard to materials, manufacturing conditions far from equilibrium can give rise to spatial bistability phenomenon (see Figure 2.6). Indeed, the possibility of spontaneously transiting from a so-called thermodynamic state to a dissipative state, controlled by the exchanged fluxes, seems to occur in ice crystals [PEL 00] or in massive graphitic carbon structures obtained by chemical vapor deposition of a pyrolyzed hydrocarbon [DEL 03]. Finally, we note that the existence of a solid generated from a chaotic state such as that observed in turbulent fluids or from nonlinear chemical reactions [DEW 09a] was theoretically discussed [RUE 82] but does not seem to have been proven experimentally. As we pointed out in Chapter 1, a local pentagonal order, like in quasi-crystals, is incompatible with an overall translation symmetry: it has been suggested that it could generate a frozen turbulent structure.

2.3.3. *Production processes*

From a practical point of view, the preparation of materials will depend on the form of energy available and the associated intensive variables. Indeed, it is necessary to control the nature of the initial phase (gas, liquid or solid) and the quantity of energy supplied or combined with a supply of matter in a closed or open system. In a gentle method, corresponding to a linear response, a small variation in temperature, pressure or chemical potential will lead to a stable or possibly metastable crystallized state: this is the example we gave to obtain a monocrystalline silicon ingot (see Figure 2.10). On the other hand, an input with excess energy causing a sudden change in intensive parameter will lead to frozen unstable phases, generally of the vitreous type. This is the case of the quenching phenomenon, with the sudden solidification of an initial fluid phase, the propagation of a shock wave or a bombardment of particles in a solid or even an uncontrolled exothermic chemical reaction. The use of an oxygen-acetylene torch to prepare carbonaceous phases is an excellent physicochemical example of these processes (see Figure 2.11). In these situations far from equilibrium, the competition between the inputs of heat or matter and the speed of chemical reactions leads to different growth mechanisms and varied morphologies for the resulting phases, with an excess of internal entropy linked to a structural disorder [VEN 84].

2.4. Conclusion

In this chapter, we have reviewed the principles of phenomenological thermodynamics by introducing the associated concept of a physically defined model system, involving a boundary or separation with an external environment. The presence or absence of exchanges of energy and matter leads to the definition of the main situations, also taking into account possible internal chemical reactions: they are isolated, closed or open systems, respectively. Apart from the states of equilibrium that allow us to define standard thermodynamic parameters, the conditions for obtaining materials are essentially by exchange of energy and possibly matter. Their formation occurs in real non-equilibrium situations in a closed or open system. The analysis, according to the intensity of the exchanges with the environment, makes it possible to distinguish two regimes, near or far from equilibrium, with the possible existence of a bifurcation point conditioning the onset of a new organization. This analysis allows us, for example, to distinguish between the kinetic processes of crystallogenesis and morphogenesis. It describes the solids formed at several scales by showing an increasing shape complexity, which are no longer just typical crystallographic polyhedra. Self-organization processes far from equilibrium lead to mimicry of living systems, a problem that will be discussed in Chapter 9. We will see in Chapter 3 that this macroscopic approach can be complemented by explaining the surface functionality and the particular role played by specific interfaces, key to system–environment exchanges.

3

Classification of Materials

An alphabetical review of conventional materials, ranging from manufacturing processes to different uses, provides a good overview of this topic [ETI 08]. Usually distributed across large families, it is not their chemical nature that will be the trickiest point but their behavior as closed or open thermodynamic systems. We will first consider the nature and the selectivity of the exchanges, permanent flows or punctual stimuli, crossing the surface or interface of a solid.

Thus, the bulk properties described in [KIT 04], a classical textbook of solid-state physics for isolated or closed systems, will be exploited independently of the chemical nature of the phase concerned. We will only distinguish mono- or polyphasic materials as constituent elements, the latter being able to form more elaborate structures. This approach is reinforced by the chronological classification presented in Figure 1.9, which distinguishes between traditional, classical and advanced materials. They are known to be empirically multifunctional, adaptive or even intelligent when they are able to improve their response according to given information. We will see that analyzing responses under different constraints and flows makes it possible to explain this classification according to the purpose envisaged. Note that the increasing importance of surfaces in nanomaterials will not be dealt with specifically in this chapter. Indeed, the quantum nature of the associated physical phenomena will only be discussed in the chapters that follow.

Classifying model systems, subject to constraints and/or specific flows, as a group of materials allows for a more rational presentation of the collected responses. An operational classification can be established: this is based on the concepts of generalized thermodynamics, including energy transformations. Indeed, the devices of which we will give some essential examples, either directly use one of the known forms of energy or operate via a conversion process (see Figure 2.3).

3.1. Role of surfaces and interfaces

3.1.1. *Nature and symmetry of a phase separation*

We find both aspects, form and functionality of a surface or an interface, in the case of an internal separation between two condensed phases. As we saw in the section 2.3 of Chapter 2, the processes of crystallogenesis and morphogenesis produce very varied forms of solids. Natural forms are not generally conserved and we have adapted them and optimized an exchange surface according to the envisaged use. On a geometrical basis, it is a symmetry-breaking operation in a crystalline system, resulting in particular physicochemical surface properties. Three main classes of surface characteristics are present:

– the topography of the surface is ideally a crystalline plane but it can be curved, rough or even nanostructured. It plays an essential role in the transfer of mechanical stresses or for an electrical or thermal flow. It may also correspond to a solid or porous texture; in the latter case, a mass supply by an adsorption and mass transfer phenomenon is possible;

– chemical or electrochemical reactivity from surface states or linked to the presence of dangling chemical bonds, conditioning an adsorption phenomenon by the interaction of Van der Waals forces or by the formation of strong chemical bonds. Low interactions give rise to atoms or small molecule migrations and superficial reconstruction to optimize surface energy. On the contrary, chemical grafting makes it possible to modify the specificity of the surface (hydrophobic versus hydrophilic);

– the presence of static electric charges with a potential barrier (asymmetric Schottky barrier) for their circulation to or from the exterior, or alternatively the presence of a direct ohmic electrical contact. In addition, recombination phenomena of charged particles can occur [KRO 01].

Thus, the shape and the nature of a surface play a key role, which is reinforced by its selectivity, usually created by a suitable treatment: this is the problem of interfacing. A typical example is deposition in the form of a thin layer for which several physical or chemical processes have been developed (as for vapor deposition, see Chapter 2). Surface treatments after the deposition of an interphase of nanometric or micrometric thickness, such as a metallization or a thin layer with a different optical index, make it possible to control an energy exchange. It is a vital parameter for optimizing electrical, thermal or mechanical contacts or to confine the propagation of an electromagnetic wave in the system. These parameters are necessary for creating a technical object or artifact that selects system–environment exchanges.

3.1.2. *Classification according to the requirements*

As we introduced at the beginning of Chapter 2, we will distinguish the static constraints from exchanges: they are all associated with intensive parameters. Constraints, which are largely mechanical, are forces exerted on the walls of the system limiting them. For exchanges, continuous flows or pulsed stimuli, we start from the thermodynamic classification using the main forms of energy available: thermal, mechanical, electrical, radiant or chemical, which were introduced in Table 2.1. The main properties of solids are thus associated with energy and mass exchanges, landmark of the material classification according to their responses.

3.1.3. *Composition of a system*

Irrespective of the type of solid, whether of inorganic or organic origin, it is necessary to schematically distinguish between structural and functional materials. The first type of physical properties corresponds to generic properties, which are essentially a function of the strength of the involved chemical bonds, of their symmetry and their spatial distribution. Upon closer observation, their textural and morphological arrangements are paramount. These are essentially the mechanical and thermal properties of all solids irrespective of their origin. The second type of physical properties corresponds to particular peculiarities related to their electronic structure, that is to say, to the types of chemical bonds involved, mainly covalent and metallic bonds. The existence, energetic distribution and occupation of electronic bands primarily define their specific properties, in essence of optical or electrical [KIT 04].

The composition of a thermodynamic system can be divided into two stages. First, as a homogeneous condensed phase of monocrystalline nature, or otherwise, as we have just discussed. Second, heterogeneous media with two or more distinct phases, which induces the existence of interfaces or internal walls providing additional specificity linked to their selectivity. A simple example is shown in Figure 3.1, showing the existence of a boundary between two distinctive phases or subsystems. They no longer exhibit uniform intensive variables (temperature, pressure, chemical potential for a concentration of a chemical species X or the presence of an electric charge). This discontinuity or presence of a gradient at the interface then implies an internal flow in the system in addition to its exchanges with the environment. For example, a boundary may be permeable or porous to certain chemical species, ranging from the boundary conditions where the concentration of X and its chemical potential are practically the same on both sides, to a step function (called the Heaviside function). In addition to the case of a controlled mass flow across a porous film, local temperature or pressure gradients are the most common. In this context, it is necessary to highlight the particular case of heat transfer, as this can be transmitted by direct contact between two solids but also by radiation or gas exchange. These modes of transfer correspond to the energy dissipation mechanisms reviewed in Chapter 2. They show that a truly selective thermal boundary in adiabatic regimes is difficult to achieve experimentally. Thus, each specific type of interface adds an additional degree of freedom to the system considered by forming a subsystem.

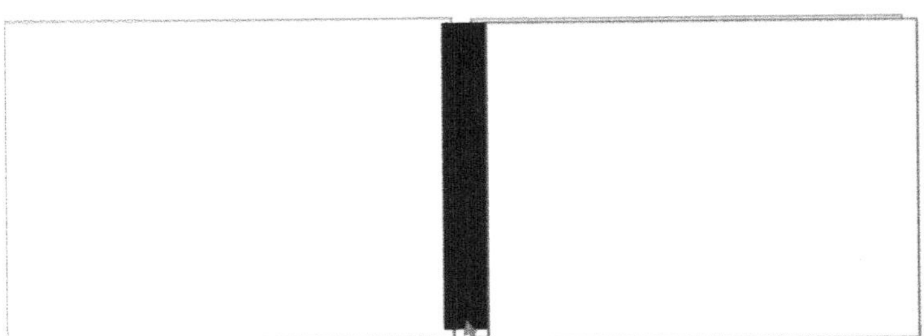

Figure 3.1. *Model of a heterogeneous system split into two interacting subsystems with a boundary selective to one or several intensive parameters*

The main categories of polyphasic materials are as follows:

– composite materials and solid dispersed media; the former are made up of long fibers acting as mechanical reinforcements while the matrix absorbs thermal or mechanical constraints and makes it possible to optimize the overall shape of the object. The dispersed media can be formed with small particles, normally randomly distributed in a continuous or matrix phase. Their properties are discontinuous functions of their volumic concentration: it is the threshold phenomenon known as percolation (see Box 9.2 in Chapter 9). The size of the dispersed particles is small; it can become nanometric and generate colloidal dispersions or solid nanocomposites;

– heterostructural units, generally based on doped semiconductors (type n or p); they form electronic components where the nature of charged interfaces, in the presence of an electrical potential gradient, plays an essential role [KRO 01];

– porous media, coexistence of a solid phase and a gas phase, and possibly liquid, which is in motion. It is convenient to distinguish between granular and cellular solids, i.e. powdered or structured solids. These are foams and aerogels with certain mechanical strength. Both are characterized by their porous volume fraction and low bulk density. The dominant factors are the distinction between open and closed porosity, pore size distribution, shape and spatial distribution. Lastly, open multiphase media must be described with two independent fluids, gas or liquid, circulating in the opposite direction, separated by an impermeable solid boundary. Each subsystem can be at a different temperature: this is the principle of heat exchange;

– a final class, even more hierarchical, is that of devices such as electric circuits or thermomechanical or thermoelectric machines. They are made up of distinct subsets comprising several modules of materials: this level of description will not be discussed in this chapter.

3.1.4. *Type of responses and functionality*

Several classes of functionalities have been developed that should be defined before proposing a classification of materials as systems. They correspond to the following main types of responses:

– the transmission of a flow or signal often of an optical or electronic nature involved in communication process. Under the influence of an exchange, a stimulus or a constraint modification, a perturbation of the response is recorded. This device is a detector or a sensor;

– transducers transform the electrical starting signal into another form of energy, e.g. actuators, where a mechanical response to the application of an electrical potential is involved;

– the conversion of a primary energy flow leading to a response in another form of energy. These converters in general transform a primary energy into electricity, a particularly useful secondary form;

– the storage of energy in all its forms: mechanical, thermal or mainly electrical. Electric energy is the trickiest to store because it is essentially an energy vector allowing transport;

– digital memories when recording a stimulus and storing an information essentially by optical, electronic or magnetic means (see Chapter 4).

3.2. Main types of materials and systems

We will classify the main properties based on information in Table 2.1. For this reason, we will examine, in each situation, the nature of the system, i.e. homogeneous or heterogeneous, and the constraints present, i.e. static forces and flows entering and exiting. In other words, we analyze the response of the system, outside thermodynamic equilibrium, which will lie in or beyond the linear response domain.

3.2.1. *Structural materials*

In the presence of mechanical forces, solid materials are designed to resist, without excessive deformation, damage by producing defects or fracture. This generic property is characteristic of all solids, such as the thermal conduction, which is associated with the vibrations of the atomic or molecular network and depends on their cohesive energy. Indeed, in a homogeneous solid, the stronger the chemical bond, the better its mechanical strength or its ability to transfer heat (Fourier's law).

Understandably all lattice limitations, e.g. the size of the microcrystals and that of the grains on a larger scale, present in a polycrystalline or polyphasic solid, are corrective factors to be considered. We will review the laws of behavior, provide a comparative example regrouping several families of materials and then identify particular effects induced or caused by mechanical stress.

3.2.1.1. *Review of behavior laws*

The general mechanical laws are based on the stress–strain relationship; three domains are defined [KIT 04] as follows:

– for low mechanical stresses, applied during tensile, compression or bending experiments, the solid deforms in proportion to the applied force. This is the elastic zone where the solid has a linear and reversible response (the limiting case being that of elastomers);

– from a stress threshold, a plastic zone appears, corresponding to a nonlinear response of the deformation. In the solid, structural defects appear, including dislocations or deficiencies at grain boundaries, with the formation of cracks that can develop and irreversibly damage the solid;

– finally, a rupture arising from the propagation of a particular crack, which depends on the type of mechanical stress and the possible presence of mechanical instabilities initiated by a cascade phenomenon that may have a chaotic trait. In a crystal lattice, it is associated with the local energy required to initiate and cause the breaking of chemical bonds.

This mechanical behavior is described by a set of experimental characteristics:

– the elastic modulus or Young's modulus E_y that corresponds to a force divided by the induced dimensional change; it is a homogeneous variable as a pressure that characterizes all the solids in the elastic zone;

– elongation at rupture under tension, associated with the accompanying load representing the maximum possible stress known as critical stress. Depending on the material, there are two types of fracture, fragile or ductile, that is to say sudden or progressive;

– the energy stored by plastic deformation reaches an optimum value at rupture: it is the maximum amount of energy or resilience upon application of an increasing force or during an impact. Different mechanisms of restitution exist mechanically or via heat production. This is, for example, the appearance of a surface energy during the creation of a crack detected by acoustic emission or by the production of heat linked to an internal friction mechanism.

3.2.1.2. *Comparison of various materials*

The current conditions of use correspond to increasingly demanding criteria. It is necessary to use homogeneous materials or composites as light as possible, resistant to many thermal variations and to the physical and chemical aggressions of the environment. Lastly, the formation and assembly techniques must be optimized to manufacture the object to make it suitable for the desired use [EVA 01].

In Figure 3.2, Young's modulus is plotted as a function of bulk density, since the search for light structures is a permanent objective in industry; an overall density of less than 2–3 is a constant aim. The main types of solids are present on this diagram, ranging from polymers and their composites, which are the lightest compared to ceramics. They include the hardest materials with a refractory nature, such as diamond, which are stable at high temperatures. Finally, metals and their alloys exhibit a high behavioral flexibility for moving parts, even though they have a higher density.

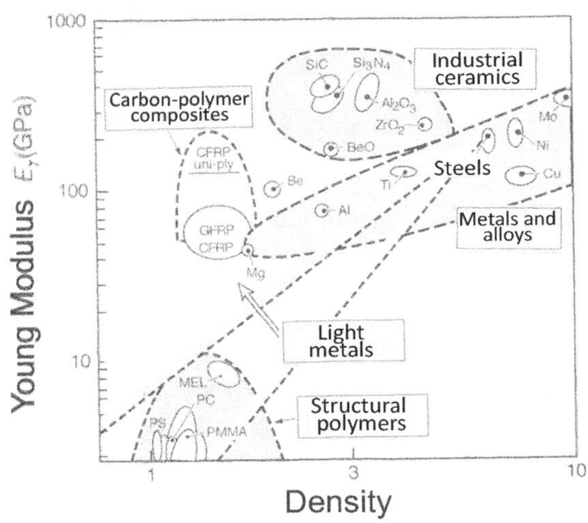

Figure 3.2. *General diagram showing the elastic modulus or Young's modulus (Ey) as a function of the actual density for different groups of materials. Representative acronyms are as follows: Carbon fiber reinforced polymer (CFRP) or graphite fiber-reinforced polymer (GFRP); concerning the various structural polymers: melamine (Mel), polystyrene (PS), polyvinyl chloride (PVC), Poly(methyl methacrylate) (PMMA)*

3.2.1.3. *Associated phenomena*

Sensors based on electrical extensometric measurements have been developed; these are strain gauges (or gages) that allow the laws of elastic behavior of a solid to be followed. They are generally constituted by piezoresistive dielectrics, such as quartz, where mechanical deformation leads to the appearance of electrical charges. More subtle effects can occur when the elastic behavior is induced by an external field, as in photosensitive polymers. These are photoelastic, electroelastic or even magnetoelastic phenomena.

Continuing this approach, also note the discovery of so-called intelligent polymers obtained from mechanical–chemical reactions that couple the two types of energy. A promising example is that of self-healing composites where cracks are filled by a polymer from microcapsules present: the reservoir is in the system [WHI 01].

Another particularly significant example is that of shape-memory materials (metal alloys or polymers) [SIL 02]. A solid changes its shape under the influence of a mechanical stress and keeps it after it has disappeared. To return to the initial shape, the solid is generally to be heated to a temperature higher than that of the permanent deformation due to the existence of a structural phase transition in the considered temperature interval.

Finally, there are chemical–mechanical instabilities corresponding to swelling phenomenon in a gel, coupled with chemical oscillations [BOI 09a]. This volume changes phenomenon, whose excitability is in a nonlinear chemical regime, as described for reaction–diffusion systems (see Box 2.1). These effects show that the mechanical stress imposed on the system can be coupled with an energy or mass flow.

3.2.2. *Electronic operators and transmitters*

Systems based on electrical energy can be considered as prototypes of increasingly elaborate thermodynamic behaviors. This is what we are going to show, starting from homogeneous materials, considering their assembly and finally by forming more elaborate electrical circuits.

3.2.2.1. *Linear responses in a monophasic material*

These properties are based on a system response proportional to the perturbation close to equilibrium. In the case of heat, electricity or material transport properties, these are Onsager's linear relations between thermodynamic forces and induced flows (see Chapter 2).

For single crystals, one must take into account the anisotropy of the physical properties, which depends on privileged directions related to the symmetry of the crystal, due to a tensorial representation, a physical magnitude connecting two vectors [NYE 61]. An example of electrical conductivity and its behavior under the influence of a magnetic field has already been defined in Figure 2.4. These simple monophasic devices can be considered as closed thermodynamic systems, exhibiting energetic exchanges with controlled heat dissipation processes in isothermal operation. Two main classes exist, metals with copper for optimal electrical transport and doped or extrinsic semiconductors with significant sensitivity to the application of a magnetic field. Their thermodynamic modeling is shown in Figure 3.3: this is the operating principle of a magnetometer based on magnetoresistance or the Hall effect. It makes it possible to control the intensity of the applied field due to a feedback loop controlling a modulation of the electric current delivered by the electromagnetic generator (as in the figure below).

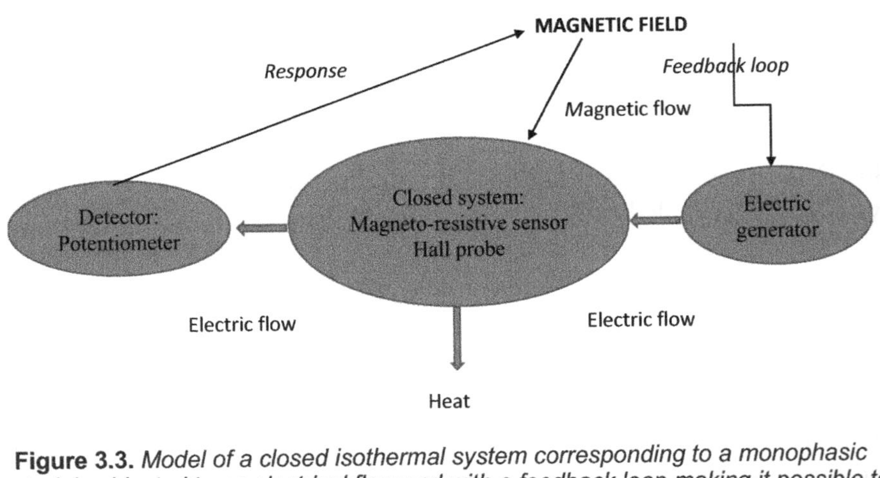

Figure 3.3. *Model of a closed isothermal system corresponding to a monophasic material subjected to an electrical flow and with a feedback loop making it possible to control the variations in the external magnetic field*

3.2.2.2. *Nonlinear effects in electronics*

We have examined the transport of electricity under an ohmic regime, but there are more elaborate devices with a voltage-current characteristic, which is no longer a linear response. These effects occur at the interface between two solids, when there is a non-ohmic contact such as a Schottky barrier [KRO 01]. Controlling these effects requires the presence of at least two phases forming the component that will constitute our thermodynamic system. This is the case of diodes with negatively or positively doped semiconductors (silicon type n or p); they exhibit a rectifying effect of the electric current that preferentially passes in one direction.

An exemplary case is that of the discovery of the transistor discovered by Bardeen and Brattain in 1947, where the density of electric current flowing in an extrinsic semiconductor is controlled by an applied voltage. Two main types are, respectively, Schottky-type bipolar transistors and field effect transistor transistors, which are the basis of current electronics. In this type of device, as shown in Figure 3.4, the current flowing between the source and the drain is controlled by a voltage applied to a third electrode, the gate. The current is at its maximum when the grid is polarized and minimum when it is zero: this is a nonlinear response in a boundary situation corresponding to an open or closed switch.

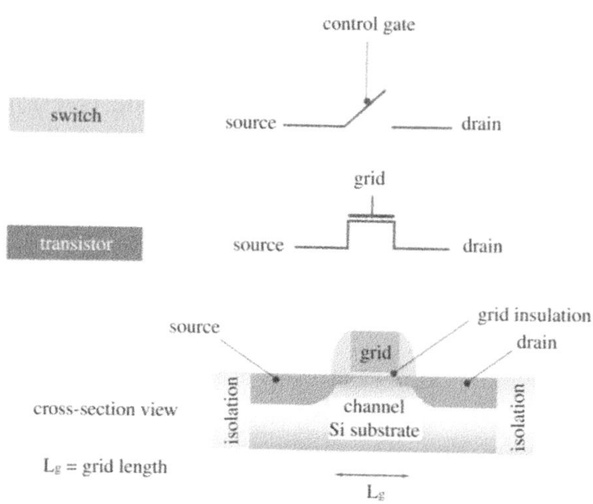

Figure 3.4. *Diagram of a metal-oxide semiconductor field effect transistor (MOSFET), which is a switch controlling the passage of an electric current from the source to the drain by means of an electrically insulated current gate, characterized by its typical length L_g; it is deposited on a pure silicon substrate that is insulating*

This device is a multiphase system based on all the integrated electronic circuits and required to carry out logic functions. For more than half a century, in order to improve the speed and the density of circuits and to reduce losses by Joule effect, their sizes were successively reduced and optimized with the development of lithographic techniques. The grid length L_g is the characteristic variable that decreases according to Moore's law and currently touches on classical physics (approximately 10 nm). We will return to this point when we will discuss the processing and storage of information.

NOTE.– Nonlinear electrical behaviors are also observed in heterogeneous media. This is the case with small conductive particles distributed in an insulating matrix forming a random lattice, i.e. in discrete percolation models. One application may be the transfer of an electric current in such a medium for use as an electric fuse for example (see Box 9.2 in Chapter 9).

3.2.2.3. *Coupled circuits and oscillators*

The combination of electrical elements leads to the formation of an electronic circuit, which has its own characteristics, as with the Van der Pol circuit (inductive coupling of a triode with an oscillating circuit). It has made it possible to demonstrate the phenomenon of self-excitation with the appearance of periodic oscillations. A sustained oscillator can be constructed with passive elements, but more complex situations are possible with the presence of non-periodic oscillations. This case was first realized by Chua in 1983 [CHU 94]. A conventional circuit comprising resistors, capacitors and inductors, with a nonlinear element such as a diode (see Figure 3.5), will have an oscillatory behavior that may become chaotic.

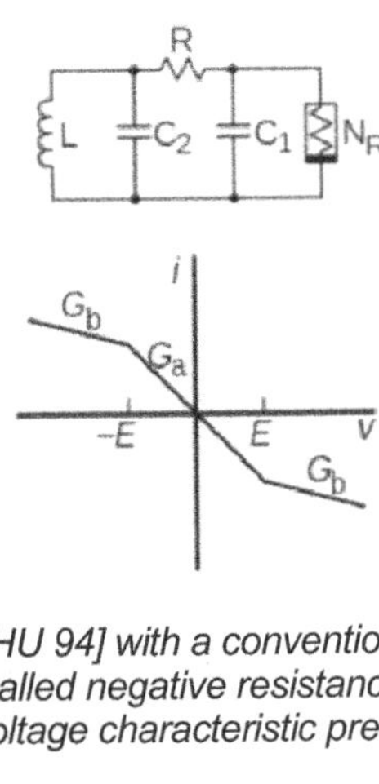

Figure 3.5. *Chua-type circuit [CHU 94] with a conventional resistor (R), an inductance (L), two capacitors (C) and a so-called negative resistance (N_R) behaving like a diode as indicated by the current-voltage characteristic presented schematically above*

This is the simplest electronic oscillator with a resonance frequency and then a frequency doubling, which then leads to a chaotic situation, as predicted by the theoretical models introduced in Chapter 2. A model of nonlinear equations gives a numerical account of this behavior dynamic. This kind of circuit has been used as an example of a paradigm for research on the onset instability at a bifurcation threshold followed by a chaotic behavior explained by the presence of an attraction basin.

3.2.2.4. *Magnetic or superconducting materials*

These two properties are associated with the quantum nature of electrons with a magnetic or spin momentum. A microscopic description is necessary to present these collective properties: this will be introduced in Chapter 5. They are often macroscopically characterized by a nonlinear response. Some solids have interacting magnetic moments that orient themselves parallel to each other: these are known as ferromagnetic. They give rise to a collective behavior that forms a magnetic domain (Weiss domain) and a permanent global magnetization in the case of ferromagnetic or ferrimagnetic solids. It can be controlled by an external action to the system and exhibits a response in the form of a rather distinct cycle of hysteresis. This is represented in Figure 3.6 by a magnetization curve, a response to a cyclic variation in the external magnetic field, which is neither linear nor reversible since it depends on the magnetic anisotropy of the crystalline solid studied. This phenomenon, common to several physical properties, is situated close to thermodynamic equilibrium. It is due to the competition between energies of volume and surface, different from a bistable state far from equilibrium (see Figure 2.7).

The presence of a mobile interface comprising the different boundaries between magnetic domains is the keys to this phenomenon. The orientation of the domains depends on the previous effects imposed on the system and must be considered as a succession of metastable states. Depending on the nature of the materials and their usage, it is convenient to distinguish between soft magnetic materials, such as ferrites, or hard (high residual magnetization) at the base of permanent magnets, for example neodymium-iron-boron alloys. Induced effects may also occur when a ferromagnetic solid that can contract or expand under the action of a magnetic field; this is known as magnetostriction. The phenomenon of magnetoelasticity, which we have already mentioned in section 3.2.2.3, is used to produce motion sensors [DU 93].

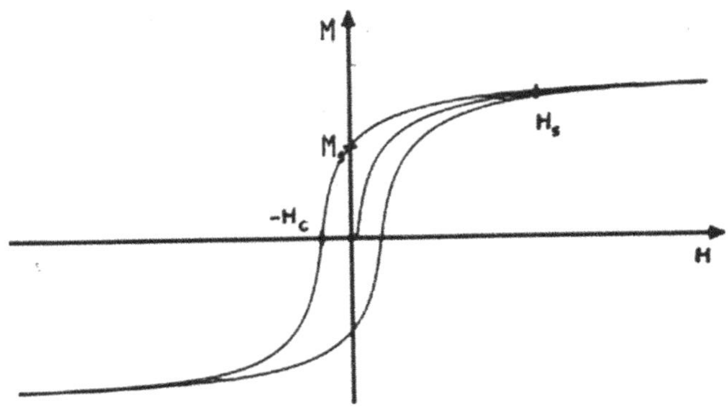

Figure 3.6. *A hysteresis cycle for a ferromagnetic material with a magnetization curve (M) as a function of the applied field (H); Hc is the residual or coercive field, Hs is the field corresponding to the saturation with complete orientation of the magnetic domains and Ms is the residual magnetization when the external magnetic field is suppressed*

Moreover, superconductivity appears only at low temperatures in some electrically conductive solids. It is characterized by a zero electrical resistance and the sudden or gradual expulsion of any magnetic field of the material [KIT 04]. It is therefore possible to envisage the transport of the electric current without loss of energy and without dissipation via the Joule effect, which is present in conventional metals such as copper or aluminum. This characteristic can be used for electrotechnical applications, essentially the production of intense magnetic fields by means of superconducting coils. The discovery of high-temperature superconductors such as ceramics, alkaline earth cuprates and other compounds has paved the way for new technological issues, although the theoretical models are still discussed [BON 05]. The applications are mainly restricted by the highest temperature of superconducting transition, a record set at 200 K in the case of hydrogen sulfide but under very high hydrostatic pressure [DRO 15]. Knowing that optimum use is at a temperature half the transition temperature, their use at the moment can only be carried out in a liquid nitrogen bath.

3.2.3. *Optical devices*

These devices are associated with a large set of phenomena based on interactions between the light and the solid, which is characterized by a complex dielectric function (see Table 2.1). It is worth considering two situations: a transparent media with only an optical index (n) and an absorption media where the absorption coefficient (k) is active. As we have seen previously, the principle of energy conservation (A + T + R =1) should be applied. This coefficient analysis versus the frequency of the electromagnetic wave characterizes the medium and its interface.

If absorption A is negligible, it is the propagation of a wave according to the laws of geometrical optics that is operational (Descartes–Snell's law at the separation surface). On the contrary, in the presence of an appreciable absorption coefficient, the Beer–Lambert law comes into play until the limiting case of a black body (A = 1). Indeed, radiative energy can then be converted into another form giving rise to various effects. These situations are represented in Figure. 3.7 for an optical system made up of a dielectric material, subjected to a light source, which acts as an energy reservoir. We will examine the main possible answers.

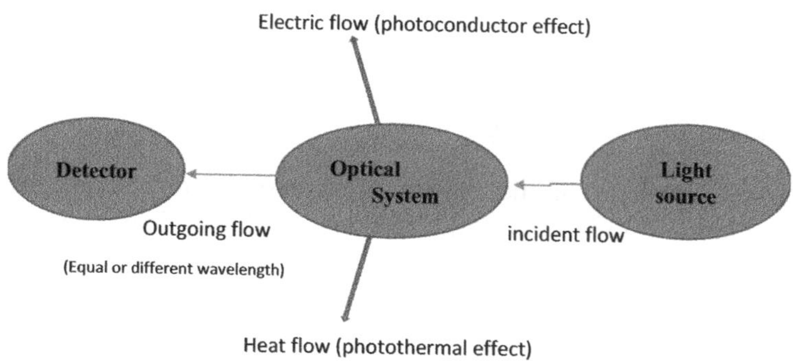

Figure 3.7. *Closed system subjected to radiative energy emitted by a light source and restored by transformation or conversion giving rise to different effects*

3.2.3.1. *Transparent media*

These materials are vitreous or crystalline in nature. For a glass, the optical index is a scalar independent of the direction of the wave with respect to the interface where transmission and reflection phenomena occur. However, for a crystal or an oriented molecular phase (crystallized polymer or liquid crystal), one must take into account the dielectric anisotropy and the optical index behaves like a tensorial variable: this is the birefringence phenomenon present in single crystals of low symmetry [NYE 61]. Normally, there is no transformation of the incident radiative energy and the outgoing flux is identical, whether monochromatic or not. Whilst many applications exist, we will only cite the most common ones, distinguishing between linear and nonlinear responses, and the main families of associated products [ZYS 91].

– Linear response to a light wave:

In classical geometrical optics, typical glass products for windows and other openings in buildings are classic examples. These are industrial flat glass, generally of inorganic origin based on silicates, but also hollow glasses used in bottles [ETI 08]. Another area in full swing is that of lenses and vision correction glasses where the glass is made up of lighter organic materials, such as polycarbonates, which are substitute for inorganic glass.

In guided optics, optical fibers are primarily used to carry a signal quickly over long distances and without attenuation. They are of micrometer diameter with a core-sheath structure that has a different optical index allowing a total reflection of the electromagnetic wave propagating along the axis of the fiber. The materials used are inorganic glass (silicas, fluorides and chalcogenides) or polymers (methyl methacrylates) [ADA 02].

– Nonlinear response to a light wave:

When the intensity of the electric field associated with the incoming electromagnetic wave increases, the response to this optical excitation is no longer proportional. New physical effects are associated with quadratic and cubic nonlinearities, which depend on the nature and crystal symmetry of the solid [NYE 61]. They are based on a modification of the frequency of the incoming wave in the dense medium or the interaction between two waves and are grouped under nonlinear optics [ZYS 91]. It is the creation of a coherent and intense monochromatic wave due to a device, discovered in the 1960s, known as a light amplification of stimulated electro-magnetic radiation (or called laser), which gave rise to this field. The discovery of new situations, whose optical bistability can range up to a chaotic state, is an essential detail. We must review the principle of operation of a solid-state laser presented schematically in Figure 3.8.

This effect uses the stimulated emission of a wave to increase its amplitude in a Fabry–Perot interferometer. It consists of two parallel mirrors treated to be partially reflective, enclosing a resonant cavity where a solid is inserted; the refractive index n of which depends on the light intensity (Figure 3.8(a)). The principle of optical amplification is to produce temporally and spatially coherent light for a given wavelength and, under certain conditions, a high emissive power. Such an optical device exhibits the bistability phenomenon far from equilibrium if, in an operating domain, two different intensities of the transmitted beam can be observed for the same value of the incident beam, known as the pump (Figure 3.8(b)). This is an example of a hysteresis loop far from thermodynamic equilibrium, such as that shown in Figure 2.7.

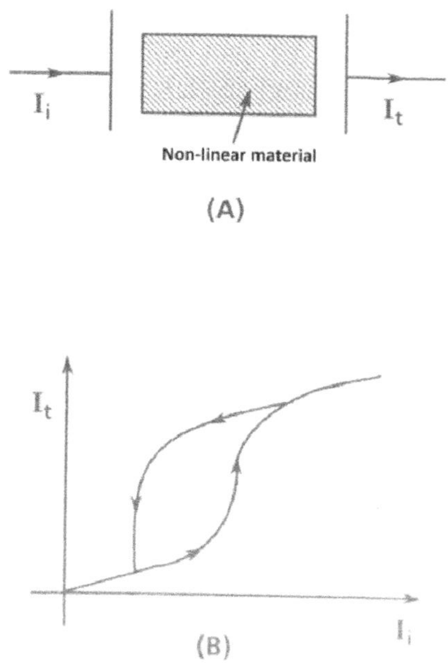

Figure 3.8. *Diagram of a laser based on a Fabry–Perot resonator containing a material whose refractive index depends on the incident intensity (Ij) (part A); this dependence (Ii/It ratio) causes a displacement of the transmission band of the resonator and induces a hysteresis effect with optical bistability (part B)*

This phenomenon is a nonlinear because the response of the solid to an external electric field, which is called macroscopic polarization, must include terms of the second and third orders as a function of its intensity [POC 02]. The polarization of the medium can produce an emission of a frequency double or triple that of the excitation frequency depending on the symmetry conditions in a crystal (in the case of non-centrosymmetric crystal groups) or for particular non-crystalline phases. Note that these optical instabilities are observed when the output power of the oscillator increases. Successive instabilities toward chaotic states have been detected, as in electrical circuits or oscillating chemical reactions [YUA 83]. The deterministic chaos in lasers is a well-studied topic due to the optimal control of the system parameters.

The materials selected are very diverse and have specific characteristics for continuous or pulsed lasers according to their characteristic emission frequency and power. These are doped insulating glass such as the Rubis laser or oriented polymers. With regard to inorganic or organic crystals, the prototypes are lithium niobate and KTP ($KTiOPO_4$) or even organic molecules like birefringent liquid crystals under an electric field [DRO 15]. With regard to dispersed media, metal colloids or meta-materials have nonlinear effects, particularly of the third order. Finally, a class of laser diodes develops particularly with heterostructural semiconductors, generating monochromatic light at a p-n junction [SAN 05] as developed in section 3.2.3.2. The choice of these materials is conditioned by their transparency in the spectral domain concerned and the efficiency of the frequency conversion. The optical quality is associated with the crystallogenesis technique in order to obtain large controlled single crystals to minimize losses due to parasitic absorptions related to structural defects.

NOTE.– In the case of a plasma, a partially ionized gaseous medium subjected to an electrical discharge, an optical bistable state was also detected. In fact, the optical emission of the plasma for a given value for electrical power can be either low or intense (dark or luminous plasma) depending on the previous situation (initiation or extinction phenomenon).

3.2.3.2. *Absorbent media*

We will consider media in which the absorption coefficient (k) will play an essential role in creating selective absorption causing a coloring and a photosensitive characteristic. A partial transformation of the incident light flux occurs, as shown in Figure 3.7. In particular, an excitation of the electrons of the semiconductor solid (photoconductive effect) or a dissipation and transformation into heat (photothermal effect) can occur. In general, the absorption of a light wave by a dye is a photoexcitation phenomenon, which triggers different transfer and relaxation mechanisms, separated into radiative or nonradiative processes. The former include photoluminescence, a particular example is that of emission of radiation from a particular form of energy (electrical, chemical or thermal). In general, this emission is separated at different wavelengths in fluorescence spectra, a simultaneous response to excitation, and in phosphorescence spectra, which has a delayed effect.

In general, this phenomenon defines inorganic luminescent materials where ions make it possible to select the emission frequency in an amorphous or crystalline matrix. These luminophores are used for illumination and display screens [FOU 02].

An alternative is that of light-emitting diodes with light generation at the p-n interface of the diode [SAN 05]. The principle of operation of a combination of positively and negatively doped semiconductors is shown in Figure 3.9.

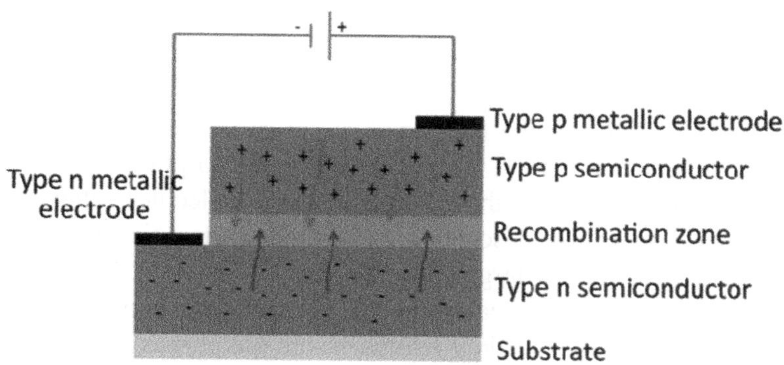

Figure 3.9. *Operation of an electroluminescent diode with a heterostructure of n- and p-doped semiconductors*

By electrically biasing a p-n junction, a recombination zone of the electrical charges creates a light emitting mechanism. The color of the light emitted depends on the forbidden bandwidth and therefore on the type of semiconductor used whether it is inorganic or organic. Diodes emitting with good efficiency an intense red or green light and more recently blue, due to arsenides and nitrides of gallium, have been created [NAK 94]. This phenomenon of electroluminescence is a direct process and therefore has a higher conversion efficiency than indirect processes. For this reason, white light emitting diodes are more efficient than other lighting systems discussed in Chapter 4.

The second are processes of non-radiative relaxation involving thermal relaxations of the crystal lattice and a production of heat. The associated absorption factor depends on the wavelength but also on the concentration of active molecules and their cross-section thickness (see Beer–Lambert law). The limiting case is that of a black body (A = 1) where the total absorption and the associated emission obey the Stephan–Boltzmann relation ($A = C^{te}.T^4$), which only depends on the temperature reached. This photothermal effect is used to store heat emitted by the sun in solar panels for hot water production or to concentrate radiation in so-called thermodynamic solar power plants. Optical switching devices with a bistable or light conversion system, as mentioned above, will be discussed in the context of optoelectronic phenomena. Thus, the phenomenon of photoconductivity, for example, applies the photovoltaic effect, which is the reverse conversion to electroluminescence, from solar energy to electricity.

3.2.4. *Adsorbers and chemical sensors*

Note that adsorption is the fixation of atoms or molecules on the surface of a solid in contact with a gas or a liquid. Desorption is the inverse phenomenon, that is to say, the release of these molecules retained at the interface. This adsorption–desorption cycle is characteristic of dispersed or porous media, providing a large available surface area and behaving as an open system with mass and heat transfers. This phenomenon is governed by the competition between the flow regime inside the porous medium and the kinetics of trapping and adsorption. In the following, we will review the different adsorption processes, the main materials used and finally the sensors developed when a specific detection is ensured [ADA 90].

3.2.4.1. *Review of adsorption mechanisms*

Depending on the nature of the molecular interactions involved, the trapped species and the nature of the substrate, we can distinguish:

– physisorption, when Van der Waal molecular interactions are low (<100 kJ/mol) between adsorbate and adsorbent of the same chemical nature: it is a reversible and exothermic phenomenon;

– chemisorption, specific and non-reversible at constant T and P, resulting from the formation of stronger chemical bonds with a higher interaction energy (from 100 to 400 kJ/mole).

In the context of surface thermodynamics (see definition in Table 2.1), it is characterized by the surface tension of a solid, which is a function of the nature of the chemical interactions present. In the case of the physical adsorption of a gas, adsorption isotherms are defined as a function of the adsorbate–adsorbent couple. Their modeling, through Langmuir molecular layers, allows the determination of the specific surface area of a porous solid, which is the accessible surface for a given molecule [ROU 09]. In general, six types of physical adsorption isotherms are recognized for these open systems. In Figure 3.10, we have reported a single example, corresponding to type 6 of the IUPAC classification, which shows the adsorption layer by layer on a homogeneous atomic plane. This is the case of a hexagonal plane of a graphite crystal, where the krypton atoms form uniform layers, characterized by the formation of successive steps at the temperature of the liquid nitrogen when the pressure increases.

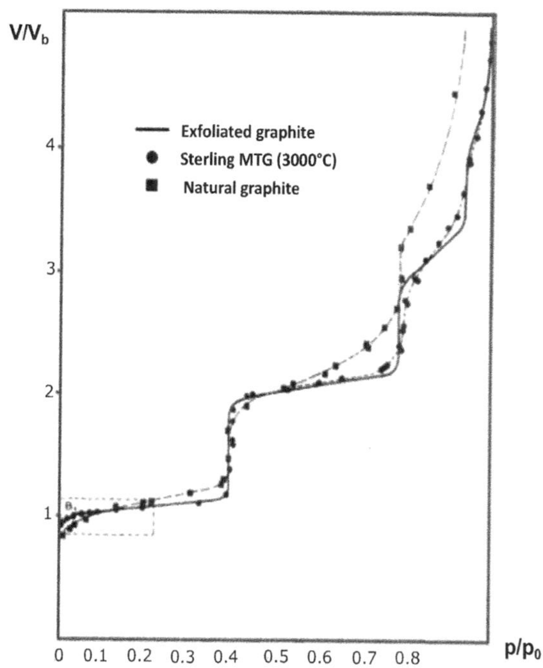

Figure 3.10. *Adsorption isotherms of krypton at 77 K where the normalized volume of the adsorbate is plotted as a function of the ratio of the relative pressures exerted; they reveal steps corresponding to the successive atomic layers. The experiments are carried out on exfoliated graphite where the graphitic planes were previously separated, on natural graphite and on a Stirling carbon black graphite (experiments of Thomy and Duval [DEL 11])*

3.2.4.2. *Porous materials*

The bulk density of granular or cellular solids is a small fraction of the actual value. They are characterized by open or closed porosities of varying sizes and shapes. According to international standards, a distinction must be made between microporosities, less than 5 μm, mesoporosities between 5 and 50 μm and macroporosities above 50 μm. Note that in solids with open porosities, their opening and size distribution as well as their connectivity are the essential geometric parameters. The overall adsorption capacity or adsorbate volume is a function of the bulk density and the pore size distribution. The process kinetics are defined by a tortuosity parameter, which allows the regulation of mass and heat transfers inside a porous medium. Finally, the selectivity of the adsorption process results from the complementarity between the suitability of the geometric pore-molecule dimensions and the chemical adhesion mechanism. An example of isotherms is given in Figure 3.11, showing a type 4 behavior in the IUPAC classification, which is characteristic of a microporous solid with a hysteresis loop close to equilibrium, due to the phenomenon known as capillary condensation [ROU 09].

Furthermore, surface chemical treatments by activation make it possible to calibrate the pores and control their hydrophobic or hydrophilic character for wetting in the case of liquids. This surface chemistry creates active sites (such as free radicals) or a particular chemical functionality to modulate the characteristics for selective fixation. The adsorbent materials are generally in the form of granules, fibers or felts filling the device in which a fluid under pressure will circulate. The main classes are activated carbons, often of natural origin [PAT 96], also silicas and porous aluminas, clays and zeolites (see Figure 1.6) or metallic foams.

3.2.4.3. *Analysis of an open porous system*

In dynamic regime, the overall phenomenon results from the competition between the adsorption kinetics and the circulation of a fluid in a porous medium (known as Darcy's law). We consider the case of a granular or cellular medium, subjected to a flow of matter in an isothermal regime. Their main use is in the field of effluent and environmental purification (molecular sieves), but other applications are similar, such as the storage of chemical energy vectors (hydrogen and methane). Heterogeneous catalysis supports or materials for liquid or gas chromatography have also been developed. In this latter process, a column of porous material, considered as a permeable fixed bed, is subjected to a multicomponent fluid moving via Darcy's law. The adsorption front comprising the gas or carrier liquid and the solute of the molecules, to be trapped in the pores, is propagated in the column until a breakthrough time when a stationary flow regime is obtained. A positive mass balance, generally associated with heat production, is obtained until saturation of the porous medium. When the maximum adsorption capacity is reached, desorption-regeneration must be carried out. In general, it is by thermal means, heating by various techniques, including the recovery of adsorbates that are often pollutants [DEL 11a]. This is the discontinuous operation of a chemical reactor.

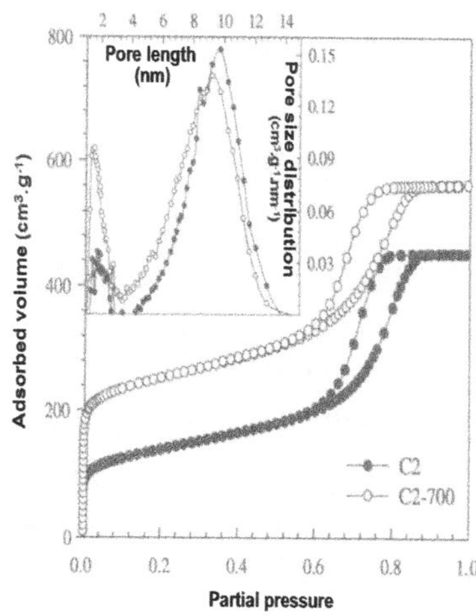

Figure 3.11. *Example of nitrogen adsorption isotherms at 77 K on a microporous-activated carbon denoted C2 (crude and treated at 700 °C); this behavior shows a hysteresis loop with an open pore size of approximately 9–10 nm, which is characteristic of the manufacturing technique (detailed in [DEL 11])*

In summary, the main sectors of activity are as follows [PAT 96]:

– purification of gases and vapors, in particular for trapping volatile organic compounds where micropores play an essential role;

– treatment of waste water or other liquids with the trapping of larger molecules that involve the mesoporosity of the environment;

– trapping of macromolecules and biological entities such as bacteria; the pore size must be even higher and the whole entity must form a macroporous solid.

3.2.4.4. *Chemical sensors*

It is a device that selectively records the trapping of a given entity by chemisorption by means of a transduction mode, allowing the creation or modification of an electrical or optical signal and its quantitative detection. These sensors are part of a very broad and evolving class of microsystems. We will cite only the main types [MIR 02]:

– gravimetric sensors often based on the phenomenon of piezoelectricity, production of an electrical polarization in crystals with no center of symmetry like quartz [NYE 61]. If such a crystal is subjected to an alternating electric field, it can resonate with a vibrational mode specific to the crystal and a variation in mass displaces this resonance frequency;

– potentiometric sensors such as an electrochemical electrode, in a solution with a membrane selective for certain species, particularly environmental contaminants [JAF 09];

– amperometric sensors capable of recording a variation in electronic or ionic current in a resistor, diode or transistor. A historical example is the use of metal oxides as semiconductors to detect certain gases. An illustration is given in Figure 3.12, where a sintered and heated tin oxide sensor is used for the detection of city gas, i.e. methane [JAF 02];

– optical detectors for the adsorption of molecular layers on a surface, which modifies the response in geometric optics; the phase and the intensity of a reflected monochromatic wave are analyzed by ellipsometry (see Langmuir–Blodgett films, Chapter 6).

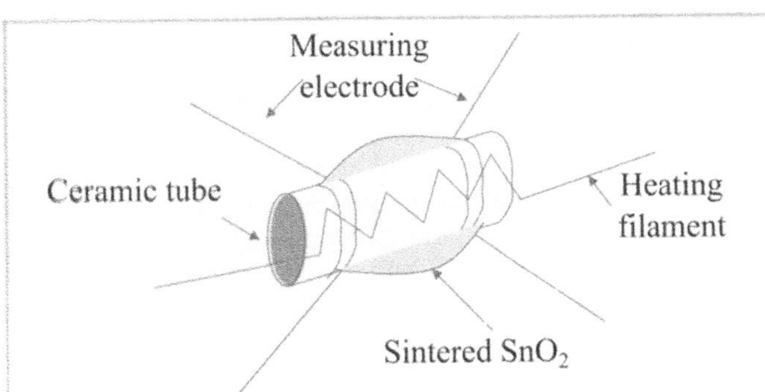

Figure 3.12. *Diagram of a sensor where the sintered SnO2 plate is deposited on a ceramic tube, including a heating filament and current measuring electrodes, which detects the modification caused by gas adsorption (according to [JAF 02])*

3.2.5. *Actuators and their analogues*

In automation, an actuator is an operational device that converts energy into work to carry out a mechanical task. We already introduced this topic in section 3.2.1.3 when describing phenomena that can cause a modification and a shape memory in a material. In this context, we are interested only in electromechanical microsystems and touch-screen devices.

3.2.5.1. *Electromechanical microsystems*

Note that electricity is the energy source causing a controlled mechanical displacement. This is the case of piezoelectric ceramics or electrostatic devices on silicon known as micro-electromagnetic devices (MEM) [MIR 02]. A possible alternative is that of electroactive and ferroelectric fluorinated polymers, which make it possible to convert the distribution of electrical charges into mechanical energy [DOM 16]. Other approaches exist just like for thermal actuators, bimetallic strip or the use of magnetostrictive elements, which we have already indicated, and finally microsystems based on optoelectronics. Applications are in the field of micromechanics with the gradual transition to a nanoelectronic scale on molecular objects: it is nanoelectromechanical devices that we will discuss later.

3.2.5.2. *Touch-screen technology*

Reverse devices for displays have been developed over the past 20–30 years. From locally applied pressure, a combination of display functions ("touch-screen" technique) is associated with a pointing and addressing device. The main technique is resistive due to a local variation in the electric field between two electrodes and to the piezoconductivity phenomenon. As we have indicated, it is a nonlinear response close to an electrical percolation threshold in a heterogeneous medium. Capacitive or inductive technologies have also been developed as well as approaches based on IR radiation. These expanding touch-screen technologies involve the processing of images and associated information, which is discussed in Chapter 4.

3.3. Conclusion

We proposed a classification according to function by examining the response of a system under various constraints and for a variable energy flow and/or a stimulus, possibly in the presence of a flow of matter. It no longer depends on the type of material involved, a common approach in solid chemistry, but on the level of response, linear or nonlinear, and the phenomena involved. This thermodynamic approach makes it possible to analyze the expected functionalities according to two complementary criteria:

– the first is to observe the difference in behavior between a mono- or multiphasic material. When several phases are present, the existence of an internal interface and specific intensive parameters for each subsystem introduces additional degrees of freedom. The two leading examples are the discoveries of the transistor and the laser during the second half of the 20th Century. This is a first step toward hierarchical devices, with more complex responses in a dynamic system that can become chaotic, such as an electrical circuit or a laser diode;

– the second point of view is to consider the concept of energy efficiency during energy detection or conversion. Maximizing yields by minimizing caloric losses and associated entropy production in sensors, transducers and other converters.

This aspect will be pursued in Chapter 4 by examining the key problems when transforming and storing energy but also in the creation of a physical support to create digital information.

4

Materials and Devices for Energy and Information

This chapter complements Chapter 3 by examining the role of materials in applications for the conversion and storage of different forms of energy. The particular case of electricity is discussed, since it is the most widely used energy in our modern societies. In addition, the current digital revolution is based on the control of the recording, storage and reading of information by developing various groups of physical memories due to the presence of a bistable-like situation. Note that information is a concept referring to the action of taking into account a fact or an event; in practice, it is a sharp stimulus, a sound or an optical or electrical signal. A signal processing technique will lead to the transmission of information by a communication process, a more comprehensive concept of information transfer and analysis.

These are the two essential points that we will discuss by classifying the technological advances in connection with conventional or innovative materials. In fact, the functionalities already presented in Chapter 3 will become tools for these new purposes. Such is the case of laser beams, diodes and transistors, but also structural materials used in various mechanical and thermal machines. This will be presented with devices that become more elaborate and must be split into subsystems to complete a thermodynamic analysis.

4.1. Conversion and storage of electrical energy

We will examine conversion and storage processes of energy in all its forms. In fact, in Chapter 3, some energy conversions were discussed with regard to sensors or transducers. Now, for a different purpose, we will also address the conversion yields and the transfer possibilities followed by storage. Two main points are discussed, on the one hand, the conversion of primary energy through electrical generators and, on the other hand, the optimized use of this electrical energy. The storage of energy under different forms is a major challenge in energy transition.

Electrical generators are generally divided into two classes. First, rotating machines in which mechanical energy is transformed into electricity: these devices comprise a gas or liquid turbine with a synchronous or asynchronous alternator. Second, fixed generators operating by direct transformation of any source of energy. The criterion of choice is the evaluation of a practical yield versus a theoretical yield (η) given by thermodynamics from the Carnot cycle. We will focus mainly on the direct transformations and the materials used by examining the main devices implemented. Then, we compare them with the transformations in successive stages, which pass through various forms of energy. In the latter case, the final yield is the product of the intermediate ones, hence is lower overall.

4.1.1. *Direct conversion electric generators*

Before discussing the photovoltaic effect, it is necessary to review the conversion of high-frequency electromagnetic radiation into electricity. Radio antennas are what convert the signal into electrical characteristics in a conductor or a transmission line. Note that the electromagnetic energy of a wave propagating in space is the flux of the Poynting vector (see definition in Table 2.1). The emission or the reception of a wave will depend on the shapes and dimensions of the antenna, which will condition the resonant frequency with a passband and a certain optimal power. Their generalization forms the basis of all current Hertzian and microwave communications systems.

4.1.1.1. *Photovoltaic cells*

This phenomenon, the creation of an electric current under the influence of light, was observed by Becquerel in the mid-19th Century. It is the photoelectric effect, corresponding to the emission of electrons in metals or semiconductors, which was interpreted by Einstein. The release of an electronic current from radiant energy gave rise to the photovoltaic effect: a schematic representation is given in Figure 4.1. It is a diode converting the radiative energy into electrical flow [MIL 07]; the opposite effect is a light-emitting diode (LED), already discussed (see Figure 3.8). Light radiation is absorbed by a diode, typically doped n and p silicon, which has an adequate forbidden energy band to recover energy from the solar emission spectrum. The interface of this heterostructure generates electrical charges, which are separated and collected, giving rise to an electric current. As for electroluminescence, the exact description of the quantum phenomenon on the microscopic scale will be introduced in Chapter 4.

The thermodynamic approach leads to a theoretical value $\eta = 0.85$ and in the case of a simple junction, the practical limit is currently estimated at about 30–40% [MIL 07]. The actual efficiency of a cell is the ratio between the electric energy ultimately produced and the light energy received. In fact, it depends on one side of the solar spectrum, its variations and conditions of capture, and on the other side the conversion rate. In addition, losses by the Joule effect are present in an operational module consisting of an inverter that transforms direct current into alternating current with an adequate potential for use.

Various semiconductor solids have been explored in recent decades [SER 13]. The silicon sector is the most developed, accounting for 98% of the production.

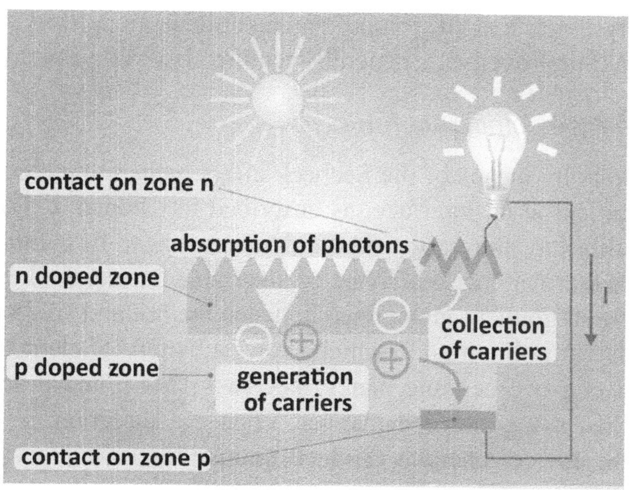

Figure 4.1. *Operation of an n and p doped silicon photovoltaic cell; under visible light there is a generation of electric charges or carriers that form a junction at the interface and give rise to an electric current I (according to [JAU 01])*

Monocrystalline Si must be distinguished from polycrystalline Si or even amorphous thin layer Si. The best cell yield for crystalline wafers is around 25–30% [JAU 01]. Other semiconductors are copper and indium diselenides and cadmium telluride, which have drawbacks regarding supply and raw material cost. Finally, more recently, the emergence of derivatives of the perovskite type gives conversion efficiencies of approximately 17–20% [GRE 14]. This material is cheap and easy to produce but its stability and aging are still a cause for concern.

Another class of materials is organic semiconductors; in particular, Gratzel cells [GRA 14], discovered in 1991, which have photosensitive pigments grafted onto a semiconductor such as titanium oxide. These photoelectrochemical cells have a current efficiency around 12%. New organic molecules (see Chapter 6) have a small yield, only a few per cent but recently stable values of approximately 10% have been achieved [BAR 16]. Advantages such as the use of a printing technique on flexible polymer supports are a promising technique; it is also possible to use new organometallic perovskites [GRA 14].

Finally, at the technological level, multijunction devices (tandem cells) or solar flux concentration devices allow yields to be improved, which can reach up to about 40% in the laboratory [GRA 14]. It is necessary to point out that the optimum conversion efficiency of an isolated cell is approximately twice that of a complete functional module in a solar farm and operational improvements generally require about 20 years to catch up.

4.1.1.2. *Thermoelectric generators*

They operate by means of the Seebeck effect, generating electricity from a thermal gradient at an interface, as described in Chapter 2. For situations close to equilibrium, they obey thermodynamic linear flux–force relations. The source of thermal energy may be of a variable nature, the key parameter being the thermal gradient maintained between the hot and cold source. Such a device is shown in Figure 4.2, which shows a battery of elements mounted in series between two ceramic plates. The two parts n and p doped silicon semiconductors in a controlled manner, impose a direction on the electric current and the device generates an electromotive force.

The energy efficiency of a thermoelectric system is calculated from the ideal Carnot cycle (see equation [2.4]) for a system with two reference temperatures; in practice, the conversion efficiency is approximately 10% [VID 94]. The search for optimized materials requires the definition and optimization of a coefficient of merit representing a high Seebeck coefficient associated with good electrical conductivity and low thermal conductivity. The optimization of this coefficient corresponds to an improved conversion efficiency using various semiconductors [SNY 08]. In addition to doped silicon, these are mainly bismuth tellurides or Si-Ge alloys, and more recently tin selenide, or stable organic compounds.

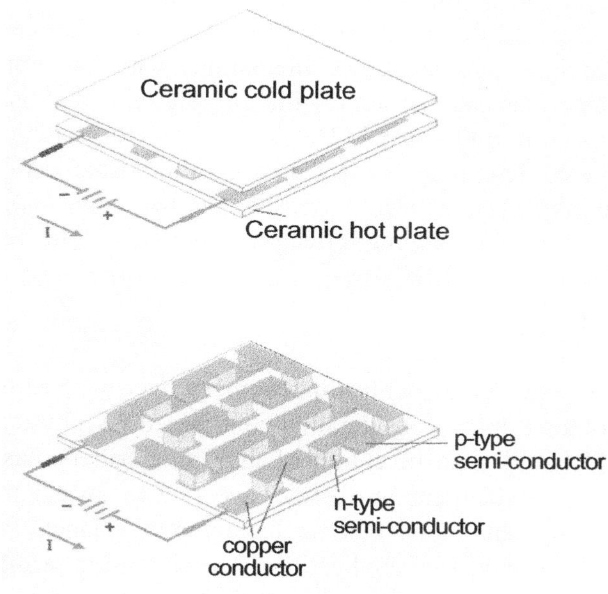

Figure 4.2. *Schematic diagram of a thermoelectric generator. An external source supplies heat to one of the ceramic plates; due to the temperature difference imposed on the ends of the pairs of doped semiconductor bars, an electromotive force is generated. In this series arrangement, the electrical potential obtained makes it possible to charge a conventional battery*

The main advantage of these devices, which can be more intricate [SNY 08], is the recovery of heat, lost during an exothermic chemical transformation, for example, which is converted into electricity. A particular example is that of thermopiles with radioisotopes, where the hot source is fed by nuclear fission; these are used in space applications despite a low conversion efficiency.

4.1.1.3. *Mechanical energy and osmotic pressure*

Among all the possible energy transformations (see Figure 2.3), mechanical energy can be transformed into electricity. Apart from conventional electric generators (see section 4.1.1.4), this is the case with marine swell due to specific transducers. The most successful results relate to the osmotic pressure for which the electrokinetic effects described in Chapter 2 are at the origin of this electrical conversion. In fact, osmosis generates a pressure difference between two aqueous solutions of different salinities through a semipermeable membrane. This is another case of modeling a heterogeneous system with an interface that is a selective membrane, as shown in Figure 3.1. The device converts an ionic current into electricity: an electric current is associated with the osmotic flow [SIR 13]. Industrial prototypes have been proposed but the conversion yields are currently very low. Developing effective nanoporous membranes is an ongoing field of research [LOG 16].

4.1.1.4. *Electrochemical cells, fuel cells and batteries*

The discovery of electrochemical phenomena over the past two centuries and their evolution have considerably contributed to physical chemistry knowledge [TRE 09]. The invention of the Volta battery has shown that reduction-oxidation chemical processes gave rise to an electrical current. Then, the development of an electric battery by Leclanché around 1850 opened the way for industrial applications. The shift from primary batteries, which can be used only once, to rechargeable secondary batteries or accumulators, was a decisive step for the storage of energy. Successive generations of accumulators since those with lead have made it possible to store greater energy densities depending on the choice of the electrode materials, the electrolyte and the potential of the redox couple. Thus, the operating principle is shown in Figure 4.3 for a lithium-ion battery [BEG 06]. It shows the specific role of each electrode as an active interface to store and restore Li^+ ions.

Much research has been carried out to increase the storage capacity by manipulating the electrodes and ions involved, such as finding an alternative to Li^+ ions (vanadium, sodium ions, etc.). Among these developments, electrolyte flow batteries have been proposed that increase the amount of energy available in a given volume [JIA 15]. Figure 4.4 shows a lithium-flow battery with reservoirs of lithium in the porous solid state separated by a polymer membrane. A liquid electrolyte is used to transport electrical charges from the reservoirs to the membrane and create a potential. A significant increase in the amount of energy is obtained yet the power delivered is lower than with conventional batteries.

From a thermodynamic point of view, a link between the quantity of heat produced and the generation of electromotive forces has been established: this is the Nernst equation, related to the free enthalpy of reaction for batteries, which are considered as closed systems. With regard to rechargeable batteries, the Faradic efficiency during a charge, which is the amount of electricity produced in relation to the quantity used during a defined time interval, is about 0.9 under appropriate conditions. Indeed, the only losses are due to the Joule effect, which depends on the characteristics of the external circuit and the power delivered. The complete charge–discharge cycle with its stability and reproducibility during cycles must be taken into account [GOO 13]. A fast discharge favors the power delivered but will decrease the efficiency. This point also raises the recurring issue of having a charge cycle as fast as the discharge cycle.

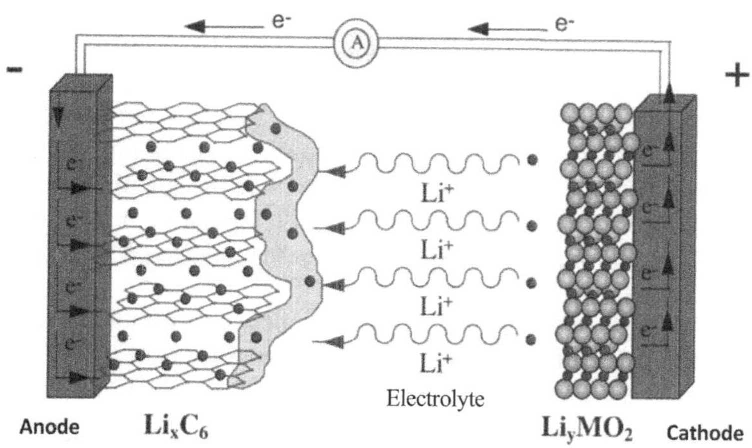

Figure 4.3. *Diagram of a standard lithium ion battery showing the insertion reactions at the electrodes with the graphite carbon anode and the metal oxide cathode (adapted from [BEG 06])*

Figure 4.4. *Flow battery using Li^+ ion technology with two reservoirs, a $LiFePO_4$ cathode and a TiO_2 anode, supplying a central electrochemical cell (adapted according to [JIA 15])*

4.1.1.5. *Electric converters, fuel cells*

In parallel, the discovery of the fuel cell in the 19th Century (by Grove in 1839) corresponds to another type of electrochemical generator. It operates based on the reverse principle of electrolysis of water, the two electrodes being separated by an electrolyte and a permeable wall. Using a catalyst, the hydrogen produced combines with the oxygen in the air to give water, heat and electricity (see Figure 4.5).

It is a heterogeneous assembly also composed of two electrodes separated by an electrolyte operating in a wide temperature range (70–1,000°C): different types of batteries have been developed according to the desired use. They are classified in order of increasing operating temperature, corresponding to higher and higher powers. There are alkaline (potash) and phosphoric acid batteries that operate at room temperature, including molten carbonates, or solid oxides operating at high temperatures [ALL 01].

From a thermodynamic point of view, a chemical reactor operates in an open system because it exchanges matter and energy. It is a cogenerator of electricity and heat; the heat released is linked to the exothermic character of the enthalpy of formation for water. Without describing these characteristics in detail, the electrical conversion efficiencies are estimated to be 0.4–0.6. It should be noted, however, that the production of hydrogen at a lower cost, which does not exist naturally on Earth, is a prerequisite. Finally, batteries based on the oxidation of methane have been also developed, but these are less efficient.

Current research is of a technological nature and aims to improve the overall efficiency of the system. It is focused on the chemical mechanisms at the electrodes, in particular concerning the optimization of metal catalysts and on the efficiency of the proton exchange membrane [SET 16].

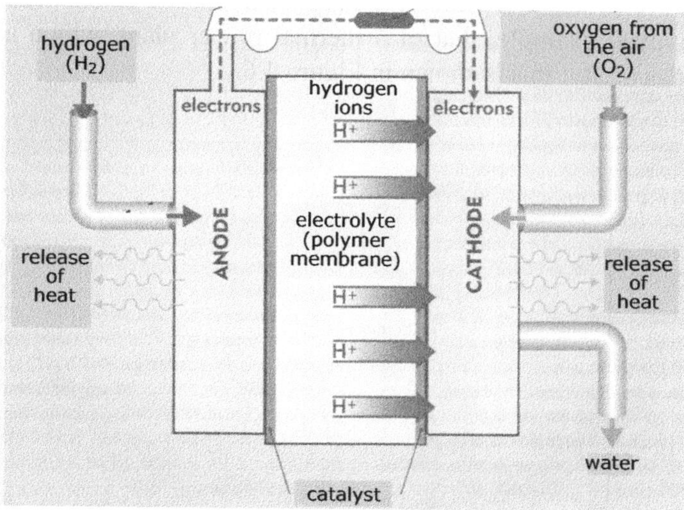

Figure 4.5. *Operation of a fuel cell powering the electric motor of a car (adapted according to [ALL 01])*

4.1.1.6. *Electric supercapacitors*

It should also be noted that electrical energy can be simply stored in electrochemical capacitors or supercapacitors using similar electrode materials. These devices comprise two polarized conducting electrodes immersed in an electrolyte. They store loads defining an electrical capacity that can be delivered instantly. These supercapacitors are of two types: double-layer capacitors operate because of the electrostatic attraction between ions and the surface of the electrodes, or electrochemical pseudo-capacitors where rapid chemical reactions take place on these electrodes. The materials used are often metal oxides based on ruthenium or indium [LOK 11] or activated porous carbons. The latter have controlled porous characteristics, making it possible to obtain a high active surface area, such as for the purification of gases and liquids and optimization of the surface density of fillers. In addition, structured nanomaterials, with an adapted active interface, present new storage mechanisms that allow a gain in supercapacitors but also for lithium batteries [SIM 09].

4.1.2. Indirect production and use of electricity

4.1.2.1. *Comparison of different types of generators*

It is a comparison between direct conversions and indirect conversions involving several steps for which the final yield must be calculated. We have chosen a typical example, that of a thermal power plant, which ultimately generates electricity: this is shown in Figure 4.6.

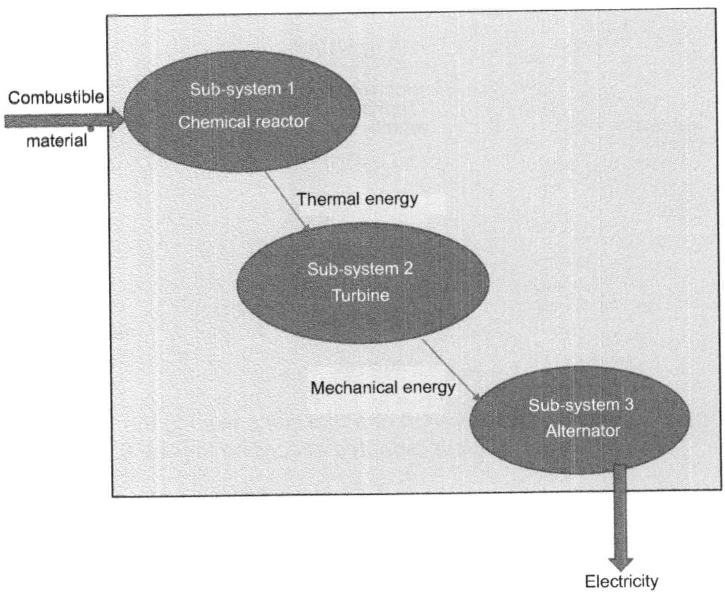

Figure 4.6. *Operating diagram of a thermal power station considered as an open system, which is divided into three subsystems corresponding to the different energy conversions*

The production of electrical energy from fossil resources occurs in three stages, each of which is considered as a subsystem with a specific conversion efficiency. The combustion of natural coal, for example, provides heat, then this thermal energy is transformed into mechanical energy in a turbine; finally, an alternator delivers electricity, which flows out of the system. The overall yield is the product of that measured for each transformation in the three successive subsystems:

$$\eta = 0.85 \times 0.40 \times 0.95 = 0.32$$

A similar situation is found in conventional nuclear reactors, known as second-generation reactors, or in high-temperature solar power plants operating by focusing the sun's rays. They lead to similar overall yields (see Table 4.1).

Other indirect transformations of energy to supply electricity are common. These include hydroelectric power stations or wind turbines, in which potential mechanical energy is transferred to kinetic energy and then to electrical energy. These thermal or mechanical machines are modular systems, generating fairly low yields; however, these are currently the main industrial devices developed on a large scale. In Table 4.1, based on the literature, we have identified the average yields obtained by comparing devices with direct and indirect energy conversions. The latter still have the advantage of superior and continuous production capacities. However, the production of electrical energy from a hydraulic dam, and therefore a natural source, stands out due to a highly efficient direct conversion.

Electric generators	Devices	Energy source	Final experimental yield
Fixed converters: – Photovoltaic – Thermoelectric – Electrochemical	 Silicon cells Seebeck generators Secondary batteries, Fuel cells	 Radiant Heat Chemical	 0.15–0.20 0.10 0.90 0.40–0.60
Thermal or mechanical machines:			
– Hydraulic dam	Turbine + alternator	Mechanical	0.70–0.90
– Wind	Blades + alternator	Mechanical	0.35
– Thermal power plant – Nuclear fission generator (generation 2)	Reactors and transformations of produced heat	Chemical Nuclear	0.30 0.30
– Solar furnace with flux concentration	Thermodynamic plant	Radiative	0.30–0.40

Table 4.1. *Comparison of the main devices belonging to the two main types of electric generators*

Improving these technologies aims to control these different processes in order to reduce the energy losses in the desired final form of energy. When analyzing the operation of these systems, it comes down to the practical concept of exergy defined in Chapter 2. This useful part of energy must be maximized by recovering part of the thermal energy dissipated by means of a thermal loop. In an optimized energy efficiency state, it is necessary to minimize the irreversible losses of the different processes by approaching the ideal conditions of thermodynamic reversibility, i.e. controlling the production of entropy in an open or closed system.

4.1.2.2. *Notes on the use of electricity*

As we have seen, electrical energy is a secondary energy that must be transferred, stored and used with the best possible yields avoiding the transformation into heat. This is the case of electric motors, which have a performance of more than 90%; this is superior to that of thermal engines, which is approximately 25–30% for cars.

Its use appears much more questionable in terms of thermodynamic efficiency for housing conditions. Without going into to the problems of heating or air conditioning, a good example is that of different forms of lighting and also display screens. Since Edison's discovery of the first incandescent lamp, electric filament lamps of carbon or tungsten have been developed. They are based on the dissipation of electric energy by Joule effect, then heating and emission of the black body. Their light output is only around 5% because they produce a lot of heat (95%) as in an electrical oven. Recent advances in this direction have been made with nanomaterials to manufacture incandescent sources that emit along a preferred axis [GRE 15].

This conversion efficiency can reach 20% in low consumption lamps, but new technologies such as halogen lamps, fluorescent lamps and those with LEDs, significantly beat this with yields of up to 50%. The light emitted is monochromatic or white with an emission spectrum covering the visible range. As discussed in Chapter 3 (see section 3.2.3), this characteristic and the associated efficiency depend on the semiconductor selected. This sparing use is of course only one aspect of the situation because the cost of manufacture and the longevity before recovery and recycling are also important parameters.

4.1.3. *Storage of energy*

The selective criterion is therefore to store a maximum quantity of energy at a given place in a form allowing a flexible further use. This is a significant issue with the development of renewable energies that are intermittent in nature. We will examine the different storage options according to the type of energy available [ODR 13], then the specific case of electricity will be approached with regard to its tricky management in a grid or mesh of a distribution network.

4.1.3.1. *General storage methods*

The first method is in the form of thermal energy in an adiabatic insulated system. It can be stored by what is known as sensible heat due to the high calorific capacity of the solid used. The principle is to recover heat coming from thermodynamic solar farms or industrial reactors, then store it in an insulated enclosure. This is the case, for example, of a natural geological repository in an underground cavity, which is never a truly adiabatic system. It is then necessary to use this heat, which must be at the highest temperature possible, to be converted with a good Carnot yield. With respect to latent heat storage, the existence of a structural phase transition with a high exothermic transition enthalpy is preferred. A possible alternative is the chemical adsorption in a porous zeolite medium. For the first option, paraffins are materials with accessible melting points or molten salts (e.g. sodium and potassium nitrates), which exhibit phase changes at high temperatures.

The second possibility is the storage of potential mechanical energy. This category includes large-scale techniques for hydraulic or compressed air storage; it also includes flywheels in engines. Mountain dams have been used for a long time but there are also pumped energy storage stations to transfer water between two basins at different altitudes. This mass of water then has a flexible future electrical production capacity with a conversion efficiency of up to 80% (see Table 4.1). The compression of a gas in a sealed enclosure or a natural cavity is an old process, which is of current interest since it can also be used in case of necessity or temporary adjustment of a production peak.

Another form of storage comes from chemical energy. Indeed, any fuel can be considered as a reserve in a chemical form. Fossil energy sources (coal, oil, natural gas) are efficient natural reservoirs, but are exhaustible stocks. The alternative for this form of energy is biomass, i.e. plants produced by photosynthesis. This natural process of storing solar energy is long but sustainable, spread over several months or years, with a very low yield of approximately 1%. Biomass can be used either directly by combustion or after chemical transformation to provide a biofuel or a biogas. Finally, the production and storage of hydrogen as a source of clean energy for fuel cells must be recognized [ART 11]. The main issue is related to its quantitative production at a reasonable cost, in particular by electrolysis of water; this will be discussed in Chapter 8.

4.1.3.2. *Storage of electricity*

Without a doubt, the greatest challenge today is to store electrical energy, manage it and release it at the appropriate time. There are two technical approaches: on the one hand, the direct capacitive storage via supercapacitors and, on the other hand, the direct capacitive storage via electrochemical systems that we have just discussed.

In order to compare these devices, a Ragone diagram is used: it allows the mass energy E to be plotted in logarithmic coordinates as a function of the specific power P (see Figure 4.7). The P/E ratio makes it possible to characterize the available power and in particular the ability of the supercapacitors to rapidly deliver significant electrical power. However, the batteries used for the massic or volumic storage of electrical energy are superior. Several generations of electrochemical pairs are shown in Figure 4.7 and show the progress made since lead batteries. Currently, lithium batteries are the most efficient (also those with the Li-S couple) and many works aim to improve their performances with regard to the storage capacity as well as tank batteries (see Figure 4.4).

To improve this capacity, as well as the charge and discharge speeds in supercapacitors, it is necessary to use electrodes made up of nanostructured materials. It is by controlling the surface-to-volume ratio of open porosities that it is possible to optimize the electrochemical storage capacity at the interfaces [SAL 14]. As we see in this figure, the faster the discharge, the farther the system is from the ideal conditions of thermodynamic reversibility; it produces heat and therefore generates entropy. On the contrary, the long-term use, up to 1,000 h, optimizes the transformation process. It is this compromise between power and energy that is sought, in particular to reduce battery charging times.

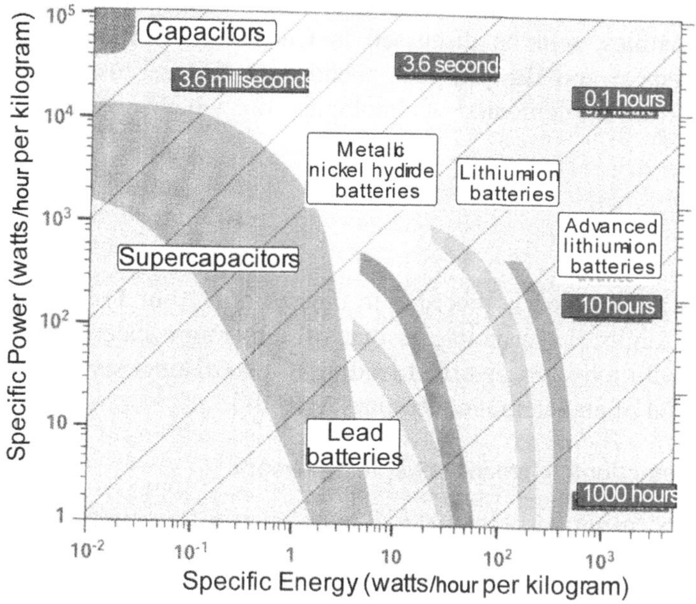

Figure 4.7. *A Ragone diagram connecting the available power (in W/kg) and the amount of energy per unit of stored mass (in W/h or J/kg) for the main electrical storage devices. We can see that supercapacitors are situated between conventional capacitors and various conventional or lithium batteries. Capacitors can deliver high power in a very short amount of time (up to 3.6 msec), but they consume very little power. On the contrary, lithium batteries have low power, with longer charge and discharge times, but they store more energy (adapted from [SAL 14])*

4.2. Recording and storing information

Analog or digital devices are used to produce and store information; only the latter will be discussed here. With the development of electronics, a physical memory, in the technical sense, is any device that records, stores and allows digitized information to be retrieved from an address box. Then, this information can be read and the signal obtained is transmitted and analyzed by a communication process.

This device is shown schematically in Figure 4.8. It is a closed thermodynamic system with a double input–output: horizontally, to record and erase the data, and vertically for the information reading and analysis. For this, we will review the general characteristics with regard to the recording and storage of digital information without explaining this concept in detail. Indeed, the theory of information, integrated into statistical thermodynamics, will be discussed in Chapter 5. Currently, it is on the material aspect and the physical phenomena that are used as memories, including the implemented technologies on which we will focus our attention.

4.2.1. *Main features*

Coded information is recorded in binary form (0 or 1) by a sequence of these two digits. Its principle is that of a memory state resulting from a response with a hysteresis effect leading to two distinct states depending on the variation of an external constraint [ANC 02].

Three functional elements exist in a memory:

– a physical support, the material to preserve information;

– a recording means, i.e. the writing range in a sensitive device;

– a restauration way or reading range corresponding to a weaker stimulus not modifying the preceding constraint (see Figure 4.8).

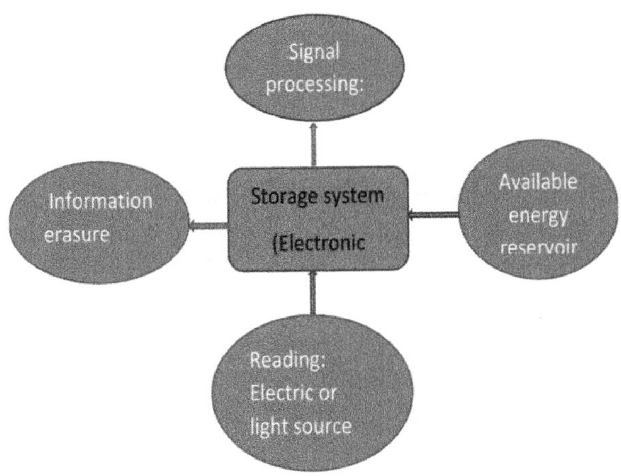

Figure 4.8. *A closed system for storing information comprising the writing and erasing process horizontally and vertically the data reading and processing device*

These memories present an ever increasing amount of information to be stored. They are characterized by their storage capacity and their access and execution speed. These two characteristics are generally incompatible and a computer system is composed of several more specific memories arranged hierarchically. Overall, we must distinguish between long-term and short-term memories; the former are read-only memories (ROMs) and the latter are random access memories (RAM), including flash memories, which can be erased in one operation. The main drawback is still the lifetime of the various physical memories, which have a large storage capacity. Research is being carried out to find universal memories, but the different types essentially depend on the materials used and on the types of energies involved in the associated intensive thermodynamic parameter (see Table 2.1). Optical and magnetic memories allow a large storage capacity for audio and video domains. Electronic memories and their derivatives favor short access times (typically ranging from microseconds to nanoseconds) for logic circuits used in computers. Thus, the transistors arranged between them allow the formation of logic gates with AND, OR, NOT functions, which form a set of operations or instructions defining an algorithm. In all situations, common issues involve the influence of the external environment (extreme conditions, chemical attacks, radiations) on the durability of storage, which are additional factors to be considered [ANC 02].

4.2.2. *Main types of memories*

The ancestors of digital memory are of mechanical origins with the Chinese abacus, the Pascal machine and, more recently, the Turing machine in the form of ribbons and punch cards. By moving from the mechanization of computational operations to programming, computing machines have become more and more powerful due to the development of memory systems that are also increasingly efficient [PER 13c]. There are currently three main types, involving radiative, magnetic, electrical or thermal energies and their combinations, all operating as closed thermodynamic systems.

4.2.2.1. *Optical memories*

The most common example is barcodes in shops or two-dimensional GR codes. Optical recording on solid supports is when information is normally written in the form of slots, usually hollow and indented and read by a differently reflected laser beam on the surface before being detected. These materials are used for recording whose starting point can be considered as analog information contained in silver photography and is based on photo- or thermosensitive polymers [BOU 84]. This is the principle of optical compact disks (CDs), as shown in Figure 4.9, which we will briefly discuss. They are composed of three or four main different layers called polymer 1, metal, a sensitive layer and polymer 2. The first polymer is often a polymethylmethacrylate or a polycarbonate while the second is a lacquer made from a urethane-acrylic resin; finally, the central zone is a weakly oxidizable noble metal.

The mechanisms involved in writing the data depend on the intended use:

– a CD-ROM contains data that cannot be modified by the user, produced by pressing a disk of polymer 1 from a matrix with slots to be read by the laser beam;

– CD-R are disks that do not contain data to begin with; it is the user who creates irreversible differences with a burner that locally burns the organic dye layer by absorbing the incident light;

– CD-RW ("rewriting") do not initially contain data; the disk can be etched by locally modifying the state of the inorganic layer from a crystalline state to an amorphous state (see phase change memories). This transformation is reversible, in particular under the influence of temperature, and the data are modifiable and erasable.

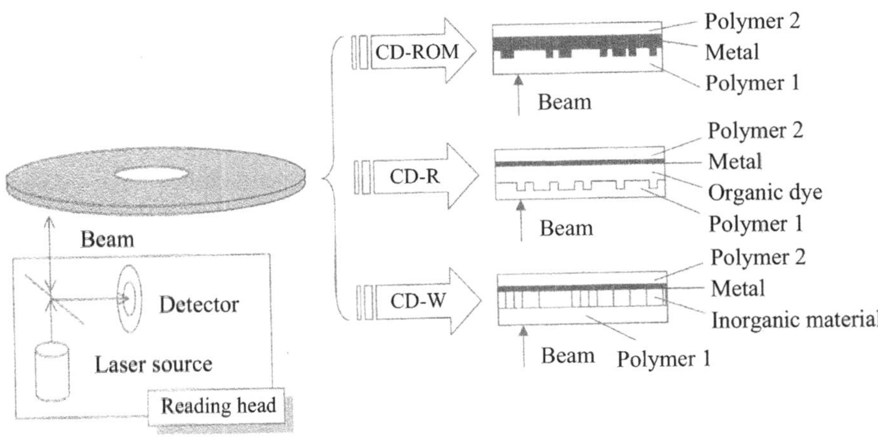

Figure 4.9. *Presentation of some digital optical disks operating with a laser beam as a reading head*

The information density is the key parameter for increasing storage capacity; it can be increased by decreasing the wavelength of the laser beam associated with the accuracy of the lithographic techniques. Thus, the CD works with a laser emitting in the infrared spectrum, DVDs ("digital video disks") in the red spectrum and the latest technology BLU-RAY with a more purple blue-violet emission. Finally, current research focuses on holographic memories based on three-dimensional laser images.

4.2.2.2. *Magnetic memories*

These are based on the principle of permanent magnetization of a thin ferro- or ferri-magnetic layer and the presence of a hysteresis loop due to the presence of moving boundaries between oriented domains (see Figure 3.6). An oriented ferromagnetic domain comprises a memory detected by the variation in electric current: this is the magnetoresistance effect (see the definition in Figure 2.4). The magnetic information depends on the size and shape of the domains, formed of magnetic bubbles for a given domain of magnetic field, which can be moved by a magnetic needle, and read using a probe. A reduction in their size leads to instability of the magnetization phenomenon and the appearance of a superparamagnetic state, which limits the available information density.

Various devices for storing digital information have been developed in the form of magnetic strips and disks as they combine random access with high storage capacity. These disks are the most developed with a technological level comparable to that of optical disks. A polycrystalline magnetic layer, consisting of grains, a set of ferrite microcrystals, for example, is deposited on a support and equipped with a reading and writing head functioning as an antenna (see Figure 4.10). An electric current passes through the magnetic layer to modify the local magnetic field and orients the domains formed. The reading is based on the magnetoresistive effect; the measured electrical potential depends on the angle between the current and the magnetization direction of the layer and it can give a binary response as a function of this orientation.

This operation can also be carried out by magneto-optic detection (i.e. the Faraday effect), but it has been significantly improved by the discovery of a negative giant magnetoresistance by Fert and Grunberg (Nobel Prize in Physics 2007). In a set of multilayers, composed of an alternating stack of iron and chromium layers, the variation in resistance as a function of the magnetic field depends on the thickness of the chromium layer (Figure 4.11). This magnetic heterostructure then exhibits a giant variation in electrical resistance (R), with a nonlinear response that is not observed in a homogeneous phase [FER 01]. This negative magnetoresistance must be compared to the classical effect on a crystalline phase, which is always positive (see Chapter 2, section 2.2.1). It is this effect that is associated with the presence of an active interface between two distinct layers, which induces this negative component.

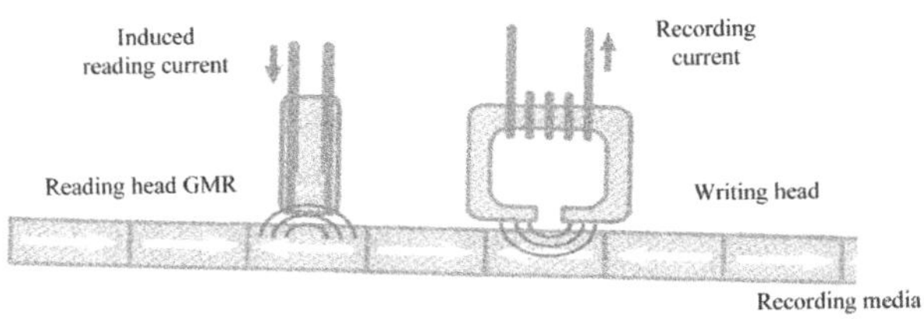

Figure 4.10. *Diagram depicting the head of a hard disk showing the writing and reading devices on oriented magnetic grains*

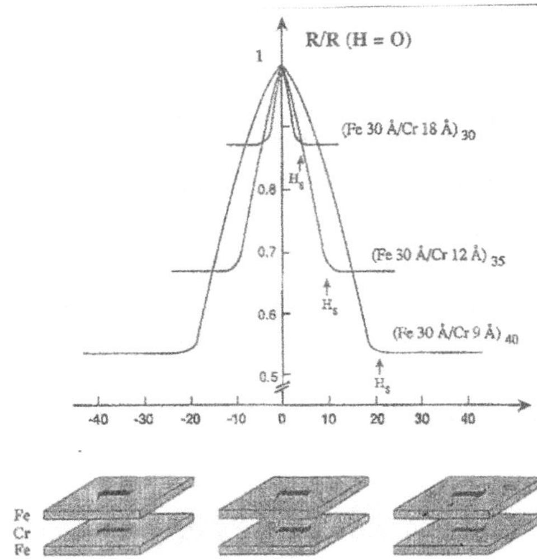

Figure 4.11. *Variation in electrical resistivity measured in the plane as a function of the applied magnetic field (H in kOe) for a series of multilayers (Fe/Cr)$_n$, where n represents the number of periods of chromium layers of varying thickness; the relative orientations of the magnetizations in the neighboring iron layers for three values of the magnetic field H are shown at the bottom of the figure (adapted according to [FER 01])*

The microscopic mechanism is a quantum effect that results from the dependence of the electron scattering in the heterostructure as a function of their magnetic moment. A very sensitive field sensor is used as the reading head of hard magnetic disks for high-density storage. We shall return to these quantum phenomena in detail in the following chapters.

4.2.2.3. *Electronic memories*

The most common technology is that of electronic memories which are easy to access but volatile. The memory consists of a transistor with a capacitance acting as a switch; in the absence of an electrical potential, the capacitor discharges and the dynamic memory is deactivated. The capacitor may be replaced by a variable electrical resistor, and thus create a non-volatile memory. Two main types of memories have been developed, either with a magnetic memory or with a phase change. The former were described in section 4.2.2.2 and the latter have a variable resistance in a phase-change semiconductor, a phenomenon already introduced for optical memories. Upon brief heating, a slightly conductive amorphous state is obtained while continuous heating allows the appearance of a much more conductive crystalline structure. This reversible conducting-insulating phenomenon, discovered by Ovshinsky half a century ago, is observed with germanium-antimony-tellurium alloys, which thus make it possible to control a transistor (Figure 4.12).

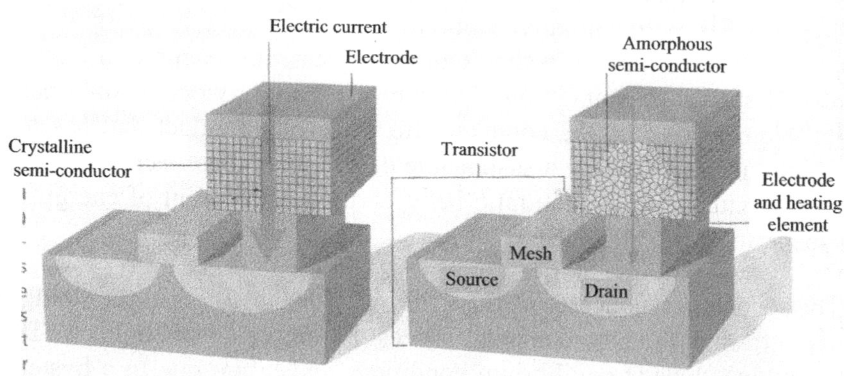

Figure 4.12. *Phase-change memories (PCRAM) composed of a transistor controlled by a variable electrical resistance in the conductive crystalline state (right) or in the amorphous state caused by the heating element and becoming insulating (left) (adapted from [PER 13c])*

Finally, another type of permanent memory that retains data when switched off is called flash memory. It is a metal oxyde semiconductor or MOS-type transistor with a floating gate where the electrons are trapped and control the state of the transistor. The discharge allows the information to be reset, which becomes rewritable, a useful characteristic particularly in nomadic applications like mobile phones. Much less energy is dissipated than in transistor memories, but as we shall see later, there are other constraints. Finally, all these memories consume significant energy, with permanent memories naturally consuming the least. The miniaturization of printed circuits is a pivotal factor for energy efficiency, but there will be a physical limit where quantum effects begin to dominate.

4.3. Conclusion

In this chapter, we have presented the different approaches for converting and storing electricity, in particular from different primary sources of energy, i.e. fossil or renewable. Indeed, all the advances aim to convert the different forms of available energies by approaching the ideal rates of thermodynamic conversion, that is to say, the theoretical conditions for the reversibility of processes. Thus, the development of mechanical and thermal machines, which operate in several stages, have fairly modest yields, which must be optimized by recovering heat for reuse and eventually transforming it into electricity. Second, electrical machines have better performances, but the key point is the quantitative storage of electricity, the progress of which depends on a great deal of technological progress, for example in transport. Finally, it should be pointed out that comparing these yields is only part of the problem since it is the complete life cycle of a product that has to be taken into account. Taking a sustainable development approach, the concept of energy return rate, i.e. the ratio between the total amount of energy used and spent, must be calculated.

The second part concerns devices and materials involved in the storage of binary information, a fundamental process in the computer domain. These devices work outside equilibrium conditions and bring, due to a hysteresis phenomenon, an increasingly efficient means of digital storage. Indeed, the creation and storage of mass data are essential elements of the current digital revolution [HIL 11]. The capacities to store, transfer and calculate are increasingly larger, a subject that we have addressed by describing the potential of existing memories. This approach leads us toward the miniaturization of the elements or parts of a nanometric system presenting phenomena, which then belong to quantum mechanics: this aspect is the subject of Chapter 5.

5

Microscopic Models and Statistical Thermodynamics

The passage from a phenomenological approach to a microscopic one is based on the corpuscular nature of matter, the presence of a set of interacting particles forming a considered system. As we have seen in the first chapter, it is the increasingly in-depth knowledge of matter that has enabled this evolution by essentially distinguishing atoms and molecules, then electrons and nuclei. A thermodynamic system is formed *a priori* of a large number of particles that must be characterized by a statistical physics approach, thereby describing the microscopic state of the system at equilibrium and then out of equilibrium. Being extended to thermodynamics, this approach allowed Boltzmann to define a statistical energy then an entropy related to the concept of disorder.

Two major domains are linked to this microscopic approach: quantum physics, which gives rules of operation such as the statistical distribution of particles in discrete energy states, and information theory, the basis of computer science and digital communication. We will first review the essential results obtained at equilibrium and then out of equilibrium as well as their consequences between the two interrelated fields.

5.1. Typical microscopic models

We will introduce the behavior of a system formed of a large number of identical and independent or weakly interacting particles initially without taking into account their quantum features. It is a random approach to events, based on an average in the phase space (combination of the space of positions and that of moments or energies). It is based on the ergodic hypothesis initially introduced by Boltzmann indicating that an average in space is equivalent to a mean value in time, which is realized experimentally. After defining the statistical entropy, we shall review the equivalence of equilibrium conditions with the state functions of phenomenological thermodynamics [CAS 70]. This will then be extended to non-equilibrium situations of statistical sets by recalling the stochastic approach.

5.1.1. *Law of distribution and definition of statistical entropy*

Let an isolated system consist of N identical and independent particles forming an ideal gas at a given temperature T. They can be found in discrete energy states E_i and their distribution on these levels is given by the Maxwell–Boltzmann statistic, i.e. a distribution proportional to exp. (E_i/k_BT). These microscopic distributions evolve towards the state of equilibrium, which is reached when they all appear with the same probability (p_i). The statistical entropy S is then defined from a probability calculation valid for a large number of particles (see Avogadro's number) [CAS 70]. The relation established by Boltzmann, which implies a relationship between an occupied fraction of the phase space and a probability, is written as:

$$S = -k_B \ \Sigma \ p_i.\text{Log} \ p_i = k_B \ \text{Log} \ \Omega \qquad [5.1]$$

where Ω is the number of possible combinations in the most probable microstate and k_B is the Boltzmann constant, a physical constant that allows identification with the phenomenological definition of entropy (k_bT is the thermal energy).

By applying the third principle, we see that the entropy is null at absolute zero for an ordered system that has reached a steady state and corresponds to the maximum entropy for an isolated system: it is a measure of the maximum disorder at any temperature T.

In a system made up of different types of particles, it behaves as an extensive variable; the overall entropy is the sum of the entropies of each part taking into account the internal degrees of freedom. This is the case with a free electron gas in a conductor or in a molecular gas. For this last situation, it will be necessary to consider separately the degrees of freedom for translation and rotation or even the mixing effect in the presence of different molecular species. The key point however is the identification of this microscopic definition with the macroscopic state function and the transition from one to the other for thermodynamic systems at equilibrium.

5.1.2. *Thermodynamic systems and canonical ensembles*

Statistical canonical units that also generalized the entropy function were introduced by Gibbs. They are derived from the distribution laws of the particles in the phase space defined by the positions and the energies of each particle. For isolated, closed and open systems, they allow identification at the thermodynamic limit with the state functions defined in Chapter 2.

5.1.2.1. *Isolated system and microcanonical ensembles*

Since the energy and the number of particles are constant in a given volume, the distribution function relies on the occupancy equiprobability of energy states. The characteristic state function is the Boltzmann entropy (equation [5.1]), which is maximum at equilibrium because it involves the greatest number of states (also see the Boltzmann H-theorem in the kinetic theory of gases).

5.1.2.2. *Closed system and canonical ensembles*

The number of particles is constant in a volume V but with a fluctuating energy since the system is connected to a constant T reservoir or thermostat. The partition function Z, established by Gibbs, leads to the definition of the free energy F as the state function:

$$F = -k_B T \text{ Log } \Sigma \exp.(E_i/k_B) = -k_B T \text{ Log } Z \qquad [5.2]$$

5.1.2.3. *Open system and grand canonical ensembles*

At constant volume, the energy and the number of particles can vary due to a specific reservoir whilst the temperature and the chemical potential (T and μ) are fixed and the particle distribution is given by the grand partition function (Q). In this example, the characteristic state function Γ is called a grand potential or Landau potential after a double summation in number and energy (the variable α introduced is linked to the chemical potential):

$$\Gamma = -k_B T \text{ Log } \Sigma' \exp.(-\alpha N).\ Z = -k_B T \text{ Log } Q \qquad [5.3]$$

These relations, in particular the minimization of the free energy function (equation [5.2]), make it possible to calculate the physical properties at thermal equilibrium, such as the calorific capacity or the magnetic properties of a homogeneous phase.

An important concept, deduced from the statistical approach, is that of fluctuations around the mean value of a variable by defining the mean square deviation in a Gaussian distribution. It is generally the energy that is considered but it can also be the number of particles; if this last fluctuation is negligible in a grand canonical ensemble, the fluctuation in energy alone makes it possible to consider it as a closed system.

5.1.3. *Situations beyond equilibrium*

Energy fluctuations play a more important role in situations under increasing external stresses or when the system is smaller in size with a finite number of particles. The transition from a discrete summation to a continuous integration, explicating the thermodynamic limit, can then be problematic. In this complex field of non-equilibrium statistical physics we will just review some significant points which are useful in the current context.

5.1.3.1. *Boltzmann transport equation*

By describing microscopic dynamics in a system close to equilibrium, it will be found to be consistent with the statistical expression of entropy. This system is formed of a large number of identical particles, independent for an ideal gas but colliding with one another. In the stationary state, by applying conservation laws, a linear resolution of this integrodifferential equation in the kinetics theory of gases makes it possible to calculate transport coefficients in the presence of a low external flux or a controlled stress. This is in agreement with the Onsager relations already defined in Chapter 2 (see Green–Kubo linear response theory).

Generally speaking, the entropy production (P), which is related to the Boltzmann transport equation, can be simply expressed from the Cercignani conjecture. This simply expresses that entropy growth is proportional to the difference between the entropy expressed by the law of distribution at equilibrium $S(\gamma)$, corresponding to a Gaussian distribution, and that of the instantaneous distribution of velocities out of equilibrium $S(f)$.

$$P = dS/dt \geq K\,[S(\gamma) - S(f)]^{1+\varepsilon} \qquad [5.4]$$

where K is an arbitrary constant. This hypothesis was explained by Villani [VIL 12] for a set of hard sphere gases in interactions satisfying the ergodic hypothesis.

In regards to the ergodic hypothesis it should be noted the temporal evolution of a system, which leads it to take all the possible configurations in the phase space, can be invalidated. The "Kolmogorov–Arnold-Moser" theorem demonstrates the existence of confined zones in the form of invariant cores for certain perturbations leading to the presence of a quasi-periodic motion that can go as far as chaotic behavior.

5.1.3.2. *Fluctuation–dissipation theory*

The mechanism of entropy production is associated with the fluctuation–dissipation theorem where the Einstein relation, relative to the Brownian motion of the particles in an isolated gas, is the first approach. The system response to a small external perturbation is assumed to be analogous to that of a spontaneous thermal fluctuation. This response is linear between a low-amplitude perturbation that measures the energy dissipation and the correlation function between two particles that accounts for the intensity of thermal fluctuations, eventually quantum fluctuations. It results in a heat-energy transformation mechanism with entropy production in a stationary steady-state closed system [CAL 85]. It can be extended to a nonlinear response when the fluctuations are no longer regressive (see section 5.1.4).

5.1.3.3. *Microscopic reversability*

Moreover, processing with statistical mechanics implies a microreversibility of time, a symmetry of the past and the future, a characteristic that contradicts the macroscopic approach. This point was analyzed in detail by Lebowitz from Boltzmann's entropy expression [LEB 13]. Indeed, the second principle expresses a principle of evolution, the growth of entropy over time. As we have already indicated, it introduces a time arrow corresponding to a process of irreversibility, which is characterized by the entropy production for an out of equilibrium situation. This paradox was also commented on by Prigogine, who analyzed the differences between microscopic and macroscopic description levels [PRI 94]. In this dynamic approach to an unstable system, it is necessary to explain probabilistic laws up to the appearance of a deterministic chaos.

5.1.3.4. *Maximum production of entropy*

As we introduced in Chapter 2, oscillating or multistable states may exist beyond a bifurcation threshold associated with the appearance of large fluctuations. One consequence is the existence of so-called dissipative structures that may even become chaotic [PRI 94]. The evolution of a system far from equilibrium is based on the principle of maximal production of entropy, which is a dynamic generalization of the second principle [MAR 06]. There is then a relation between the production of entropy and an out of equilibrium statistical distribution such as that formulated by equation [5.4]. The evolution of a system is based at the microscopic level on the optimal variation of a distribution of particles, characterized by a position and velocity in the phase space, reaching to the most probable state. By taking a dynamic approach, the system selects their path of evolution or trajectory among those available, minimizing the potential and associated energies, while maximizing its entropy that is evolving at the maximum speed under the imposed constraints [GRA 08]. This aspect, including the meaning of an irreversible evolution of the system, will be resumed in the context of the theory of information.

5.1.4. *Stochastic thermodynamics*

In recent years, a stochastic approach in relation to statistical mechanics has been developed in order to extend the concepts of classical thermodynamics to the microscopic level by considering the individual trajectories of particles in a dynamic system far from equilibrium [SEI 12]. This theory is applicable to small systems containing few particles by generalizing the fluctuation–dissipation theorem and by expressing a stochastic energy and a production of entropy. This microscopic approach is based on the dynamic description of a Markovian process where there is no influence from the system's previous states during evolution. Using the Langevin relation on Brownian motion and the classical Fokker–Planck equation allows essential thermodynamic quantities to be calculated. This is a development of the considerations on irreversibility [LEB 13, PRI 94] with a dynamic generalization of the microscopic description. It is even extended to nonlinear systems which can present a memory effect, that is to say an influence from previous situations on the state of the system.

5.2. Quantum statistics

We shall review the quantum characteristics of particles, in particular the electron which has a spin, the intrinsic kinetic moment at the origin of magnetism. Its influence on the laws of statistical distributions will lead us to a thorough description of physical properties. For this, we will start from a system composed of identical, indistinguishable and independent particles and then examine more real-world interactive situations [KIT 63].

5.2.1. *Review of concepts*

The indiscernibility of a particle is linked to the Heisenberg uncertainty principle and to the property of the representative wave function (see Schrödinger's equation), i.e. symmetric or antisymmetric with respect to their exchange in a pair. This principle indicates that any quantum object is perturbed by an observation: this perturbation is also related to the error in the physical measurement itself. Note that this approach is based on the wave-particle duality, initiated by Planck with the photon and expressed by de Broglie's equation. This duality is essential in quantum mechanics: it involves the concept of the superposition of states with the notion of entanglement between particles, fundamental characteristics in quantum electronics.

If a quantum particle corresponds to a so-called antisymmetric wave function, it obeys the Pauli exclusion principle: a single particle present per energy state. Generalizing this result shows that there are two types of quantum particles, also characterized by a mass and a charge, as a function of the quantification of the spin: a spin can be a half-integer, as for electrons ($s = 1/2$), or a whole integer zero or 1. The first group, known as fermions, are unique in a given energy state due to their spin, while the latter, bosons, can be present in any number. This implies occupation probabilities and expressions of very different distribution functions, called the Fermi–Dirac and Bose–Einstein functions respectively.

5.2.2. *Quantum distribution laws*

For a given set of N particles in a given volume at a constant temperature T occupying energy states E_i, it is necessary to calculate the law of distribution of the particles in these quantum states, taking into account their possible degree of degeneration, and a large partition function (equation [5.3]). This calculation leads to two distribution laws f(E), which we will compare [CAS 70].

5.2.2.1. *Fermi–Dirac function*

This statistical calculation shows that the mean distribution between the different energy states or quantum distribution functions is given by the following formula:

$$f(E) = \{\exp.(E - \mu)/k_B T \pm 1\}^{-1} \qquad [5.5]$$

In the Fermi–Dirac statistic, the denominator must read +1 and in this expression μ is the electronic chemical potential, positive or negative, which is called the Fermi level. It represents the last state occupied at absolute zero where the binary occupancy number changes from 0 to 1. For a positive temperature, the distribution function is smoothed around the Fermi level where the occupancy probability becomes ½. It governs, for example, the electronic conduction in metals and semiconductors as a function of the filling rate of an energy band.

5.2.2.2. *Bose–Einstein function*

The distribution function is then written with the term –1 in the denominator of equation [5.5]. In this case, the average occupancy number, a positive variable, diverges when the energy becomes equal to the value of the electronic chemical potential μ, which can only be positive or zero. If the number of bosons in the system is variable considering the grand canonical distribution, the chemical potential is zero and equation [5.5] is simplified: it is the Planck distribution used for photons or phonons that will be introduced in section 5.2.2.3 (see Table 5.1).

An important consequence observed at low temperature is the accumulation of almost all the particles in the fundamental quantum state. This behavior, known as Bose–Einstein condensation, was observed in very cold gases (Rubidium) and assemblies of excitons. It is the basis of the phenomenon of superconductivity, the transport of electric current without dissipation of electrical energy. Indeed, at low temperatures, electron pairs, known as Cooper pairs, are constituted (s = 0) and become bosons that can be collected in the fundamental state of the system (these quasi-particles are explained in Table 5.1).

Finally, we must consider the limiting case where the value is exp. (E – μ) >> |1|, which becomes negligible in the denominator. For a so-called degenerate gas, few particles are present in relation to the number of available energy states. In such cases the classical Maxwell–Boltzmann statistic is a good approximation.

5.2.2.3. *Validity limits of a Gaussian distribution*

In a system where N is large the theory of probabilities shows that an energy distribution of independent events in a gas of particles in thermal equilibrium will follow a bell curve. This is Gaussian distribution, also known as the normal distribution law, giving an average value with a defined standard deviation (see the central limit theorem). However, for nanosystems containing few particles or in an unbalanced state in the presence of external events, this law of large numbers is not always adequate. Other statistical tools, such as Lévy's stable laws for example, have larger distribution tails (power laws instead of exponential decay). They are effective, for example, in quantum optics at very low temperatures [BAR 01]. When extreme situations outweigh the average behavior of independent contributions, this approach is more realistic but it challenges the statistical definition of thermodynamic state functions. This is the case in some physical systems but also, as we shall see, in biology and economics.

5.2.3. *Elementary excitations and quantum particles in solids*

An elementary excitation is the smallest quantified excitation present in a system, a concept introduced by Planck with the phonon. It associates a particle or a quantum quasi-particle with a vibratory wave [CAS 70]. This is the concept of wave–particle duality and its generalization in field theory.

As we have shown, statistics are valid for particles without interactions, that is for a perfect quantum gas. In the case of a metal, this amounts to considering a free electron gas independent of its quantified vibrations or phonons and the atomic lattice that supports it. In the presence of relatively weak interactions with the environment, a quasi-particle will take this into account by keeping a linear superposition of the individual characteristics and the same behavior law as in a real gas. The typical example is that of delocalized electrons, which will become charge carriers in bands of energy, electrons or holes according to their charge (e) of the opposite sign; they also have an effective mass (m*) different from the real mass while retaining a

½ spin characteristic of a fermion. Note that in the case where interactions between particles become significant, the approximation of a gas is no longer sufficient; we must consider a quantum liquid with an adequate mathematical formalism [KIT 63].

To summarize the situation concerning the principal elementary excitations in solids, we have grouped them in Table 5.1. Their main characteristics are classified according to the type of energy and the type of properties involved. In this way, we propose a microscopic classification similar to that of Table 2.1, established at the macroscopic level.

Energy type	Waves and fields	Quantum particles	Physical properties
Thermal	Elastic wave	– Phonon: massless Boson, variable number	Internal energy and specific heat
Radiative	Electromagnetic wave	– Photon: massless Boson, variable number	Optical properties
Electrical	Electric field (ultimately magnetic field)	– Charge carriers: (s = 1/2, m* > 0, +/- e) Fermion	Electronic and transport properties
		– Electron pairs: (s=0, 2m*, 2e) Boson	Supraconductivity
		– Electron-hole pair: exciton (s= 0 ou 1, 2m*, 0) Boson	Optoelectronic properties
Magnetic	Wave spin	– Magnons: (s = 1) Boson	Ferromagnetism and derivative states

Table 5.1. *Examples of quantum particles and elementary excitations or quasi-particles in relation to the type of energy involved and the main physical properties resulting in solids (adapted from [KIT 63])*

In Table 5.1 we find that the nature of the quantum particles is fundamental for the description of the properties and, more surprising, that they can change their characteristics according to the applied constraints. This phenomenon depends on the intensive variables imposed on a closed or open system such as the temperature. In addition, this quantum description is essential for describing nanoscale properties, particularly when a statistical approach do not apply due to too small a number of particles. In this situation, it is necessary to take into account the discrete character of the energy levels and the additional phenomenon of quantum fluctuations linked to the principle of uncertainty.

5.3. Information theory

The essential results of this theory were established during the second half of the 20th Century. We will indicate in particular the links with statistical thermodynamics by the inclusion of the concept of entropy and associated energy cost in the creation and erasure of energy. This approach is a consequence of the principle of maximal production of entropy in systems far from the equilibrium of which a historical description makes it possible to understand the recent evolution [SEI 12]. This approach, as connected to classical as well as quantum statistical physics, will be discussed after reviewing the concepts of information and communication.

5.3.1. *Shannon–Brillouin model*

There is a general consensus that Shannon and Weaver's book, published in 1949, is the starting point of a new scientific field, a mathematical theory of telecommunication or information transport between a source and a receiver involving a quantification of this information [MAR 09]. This concept includes the creation, transmission, storage and reading as necessary technical operations of the overall process (see Figure 4.8).

To define the quantity of information, Shannon sought to quantify the choice or uncertainty of an event from a probability formula where p_i is the ith operation in a succession of independent events:

$$H = -k \Sigma p_i \log p_i \qquad [5.6]$$

where k is an arbitrary positive constant and the logarithm used is decimal but the analogy with the expression of statistical entropy (equation [5.1]) is clear. The quantity H, called the informational entropy, accounts for the missing information in a process. This function is maximum if all events are equally likely and decreases with successive selections. In practice, a binary message is considered coded 0 or 1, a practical coding for digital machines based on an experimental open-locked bistable state device which we have presented at the end of Chapter 4. This makes it possible to define a bit ("binary digit") and the amount of information in a message, sequence of bits and logical operations in Boolean algebra.

The links between information and entropy have been widely discussed by introducing the concept of negentropy. The equivalence between the two relations was studied by Szilard by imagining the intervention of the Maxwell demon capable of sorting the particles by thought according to their kinetic energy between two connected gas chambers [MAR 09]. Then, Brillouin emphasizes the need for the demon to acquire information [BRI 59], an argument taken up in particular by Bennett. Negentropy is a negative entropy, the amount of information missing in the system that can only diminish. Brillouin also states a generalized Carnot principle: the sum of this negentropy and information can only diminish in an isolated system. It also states that there is a minimal energy cost for information during an irreversible transformation. These considerations extended to systems interacting with the environment led to the definition of information as a physical quantity [SEG 13], leaving aside the cognitive aspect that will depend on the man–machine interface (see section 5.3.4).

5.3.2. *Energy and information: the Landauer principle*

The possibility that information presents a physical nature and can be converted into energy has been discussed by several authors since its formalization by Landauer in 1961 [LAN 61]. This author established that an irreversible logic function requires minimal dissipation of heat. The proof of this transformation has been validated by a thought experiment based on the use of the Maxwell demon proposed by Szilard [MAR 09], which is based on the equality of Jarzynski, who established a relation in statistical mechanics between the non-equilibrium average work performed and the free energy difference between two equilibrium states described in a canonical ensemble [TOY 10]. This result is part of the stochastic thermodynamics introduced previously [SEI 12]. In this context, an additional step was taken with the practical realization of an autonomous Maxwell demon [KOS 15]. The considered system exchanges with an electronic or mechanical device, which extracts information by reducing its entropy and its temperature.

Initially, Landauer assumed an equivalence between Shannon's information based on a distribution of logical states and the statistical entropy of a physical system represented by its microscopic states. The erasure of one bit of information requires a minimum energy spend equal to k_B T. Log 2: this is the Landauer limit or barrier [LAN 61]. Thus, any irreversible logical operation such as the suppression of a bit gives off heat since it is a dissipative process. This extremely low energy (approximately 3×10^{-21} Joules at room temperature) was measured by a special experimental device (see Figure 5.1) [BER 12]. Two micrometers of silica ball plays the role of a bit, immersed in liquid and immobilized by an optical clamp created by a laser. A second identical optical trap is created next to it and the microscopic bead can pass from one to the other; it is the equivalent of a bit with values of 0 or 1. By creating a small flow of liquid, the colloidal particle can modify potential wells and the measured average energy required corresponds to the calculated variable, the Landauer limit.

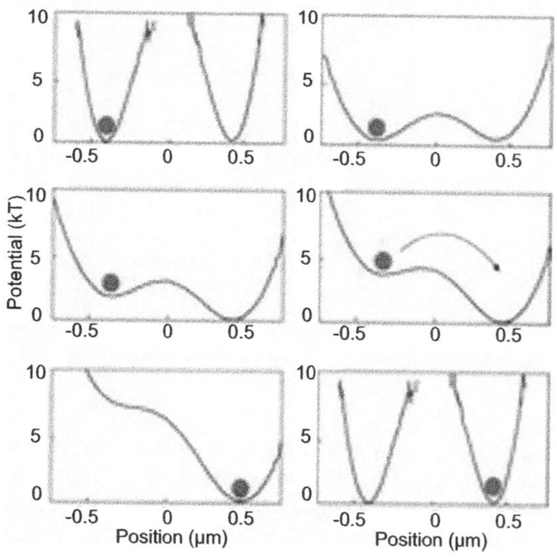

Figure 5.1. *Model of a double well separated by a central potential barrier; the successive figures represent the transition from the initial state a to the final state f by showing the deformation of the potential curves detected by a fast camera (according to [BER 12])*

This experimental test of the Landauer principle has been confirmed experimentally by anisotropic magnetic nanomemories [HON 16]. A graph shows that real cases dissipate less and less energy, about 3 orders of magnitude higher, but that their evolution (Koomey's empirical law) would make it possible to reach the theoretical limit after 2030 (see Figure 5.2). The energy expended in an increasingly small device should reduce this factor while maintaining a reliable informative character, thus approaching the theoretical barrier. The alternative way is to design a reversible calculation that is free of this limit but is not operational [DRE 12]. Thus, these results confirm the relationship between information theory and statistical thermodynamics.

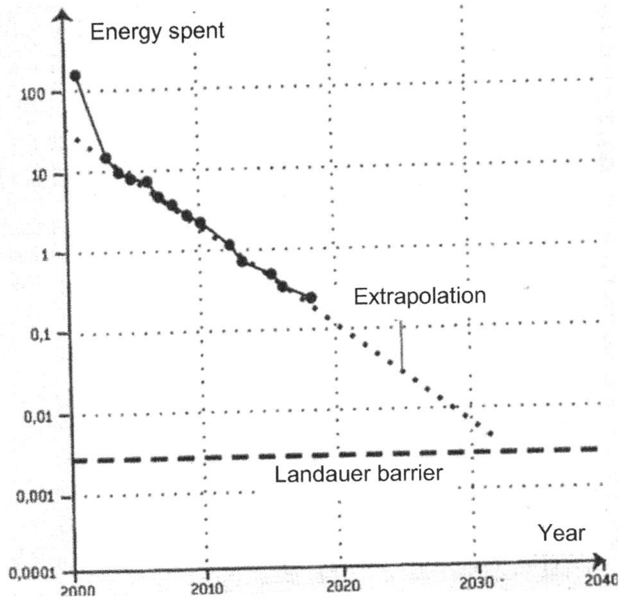

Figure 5.2. *Semilogarithmic evolution in recent years of the energy spent for an irreversible calculation operation in complementary metal oxide semiconductor (CMOS) circuits; the variation of energy dissipated was standardized in 2012 allowing the Landauer barrier to be positioned (adapted according to [DRE 12])*

5.3.3. *The role of quantum mechanics*

In systems of reduced size or operating at very low temperatures, quantum behaviors are essential. This is the case for an electronic particle in a well or a box. A quantum well is a nanoscale device such as that shown in

Figure 5.3 and a box will consist of a succession of wells and barriers with quantified energy states in each of them. From the postulates of quantum mechanics (see section 5.2.1) the following three phenomena are essential:

– quantum energy at the zero point or confinement energy, linked to the principle of uncertainty;

– Coulomb repulsion that limits the addition or the extraction of an electron from a quantum well known as a Coulomb blockade (see right-hand side of Figure 5.3);

– the tunnel effect, linked to a quantum fluctuation, through the insulating barrier and controlled by the spin value; it can occur with charge carriers, i.e. electrons or holes, which are fermions. With electron pairs forming a boson (see Table 5.1) they constitute the origin of the Josephson effect in superconductivity.

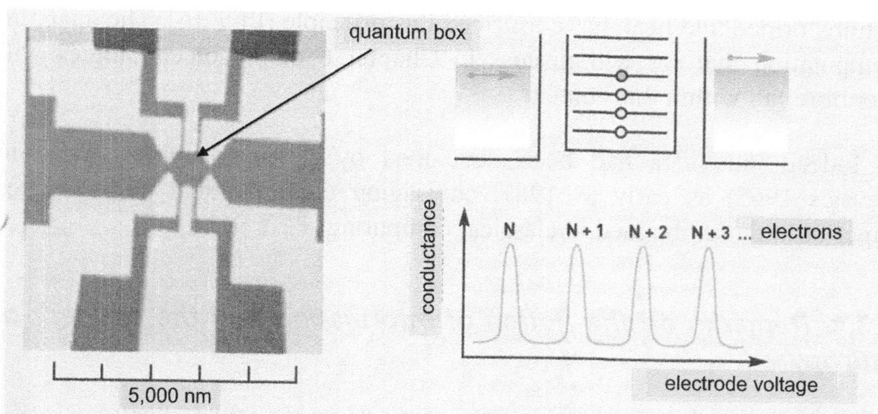

Figure 5.3. *Example of a quantum box (left) that is a nanometric island semiconductor of electrons connected by electrical contacts that are tunnel barriers; the discrete energy levels (right) due to the confinement of the small number of electrons and the observation of conductance peaks as a function of the lateral gate voltage associated with the tunnel effect showing the quantification of the charge (according to [GLA 05])*

In this context, nanoelectronics will be dominated by these behaviors associated with a quantification of electronic states. In the wave–particle duality, the wave-like aspect of the charge carriers becomes predominant and a conductance is defined as the sum of the transmission coefficients of a propagating wave multiplied by a quantum of charge. It is given by a Landauer formula that will govern the phenomenon of electronic transport called ballistics [IMR 99]. This regime is characterized by inelastic collisions occurring only at the edges of the sample, which happens for nanoobjects and molecular assemblies.

In addition, the storage of information will be defined by a quantum analog of the bit, the qubit. In a quantum system with two states, 0 and 1, unlike a conventional bit, a qubit can be both 0 and 1 but also present intermediate states (see Bell's formalism in quantum mechanics). Note that the statistical entropy is then defined by Von Neumann's expression using the technique of the density matrix [SCH 10]. This situation, due to the superposition of the quantum states during the coherence time of the system, induces both limitations and advantages for their detection and use. Thus, the demonstration of the Landauer limit in a quantum system, where information is transformed into heat, has confirmed this principle [PET 16]. The quantum computation that we will discuss in Chapter 6 on nanotechnologies will therefore fall within this context.

Indeed, this idea had been developed by Feymann (Nobel Prize in Physics 1965) as early as 1981, envisaging computational performances superior to those obtained in classical computing [FEY 96].

5.3.4. *Remarks on the notion of information and the concept of entropy*

5.3.4.1. *Equivalence between information, energy and entropy*

In Chapter 2, we introduced the phenomenological concepts of energy and entropy for equilibrium or nearby states. From work by Clausius carried out by Gibbs and then Duhem in particular (see equations [2.1] and [2.5]), we know that the entropy produced is associated with uncompensated heat generated during an irreversible process. At the microscopic level the statistical entropy, defined from the Boltzmann relation, is introduced by a constant, which has the physical dimensions of phenomenological entropy.

Thus, the equivalence between the macroscopic and microscopic levels has been recently discussed and confirmed on axiomatic bases [WEI 16].

Current developments show that there are two key points:

– in an irreversible process, the only one that works experimentally [DRE 12], erasing a bit is a dissipative process that can be considered as the release of an energy stored on a physical support;

– the inscription can be carried out by converting the heat into information and work, supposedly counter to the second principle, by materializing the role of Maxwell's demon which provides information [LU 14].

In summary, non-equilibrium statistical thermodynamics can be applied in classical and quantum regimes by considering the whole observer–system–environment set. They confirm the close link between information and thermodynamics with a concept of generalized entropy. Thus, the concept of transferred information flow can assume physical significance as for energy or matter inside a unified approach [SEG 13].

5.3.4.2. *Link with maximum entropy production*

As we have seen, the system is then governed by the principle of maximal production of entropy that we have already introduced [GRA 08]. Its formulation in a dynamic regime belongs to non-equilibrium statistical mechanics and has been the subject of several theoretical studies; it is a variational principle that makes it possible to predict the microscopic evolution of a dynamic system [SCH 10]. Note that in this context, the founding works of Jaynes [JAY 57] showed that statistical probabilities are representations of the information available for a given system. It is characteristic of our level of knowledge in an evolutive system rather than the expression of the inherent probability of an event. It is a subjective approach for an observer and no longer objective or only random for the formulation of maximum entropy. It is then calculated by the algorithms MaxEnt and MaxCal ("maximum entropy" and "maximum caliber") based on the partition functions of the system considered [PRE 13]. The statistical inference method is then based on the Bayes rule that derives an inverse probability applying to the probable cause of a given effect, as a consequence of previous facts or known information. The re-evaluation of hypotheses based on a "weight" of evidence is then a first step toward the concept of learning and acquisition for intelligent systems in a cognitive process.

Thus, the principle of maximal production of entropy in an open system far from equilibrium is a guide for modeling the evolution of matter, energy and information flows. It applies in biology and can be extended to ecosystems but also in economics and sociology [GRA 08, PRE 13]. These points will be discussed in Chapters 9 and 10.

5.4. Conclusion

The microscopic statistical description brings new light to thermodynamics. This approach was introduced by Boltzmann with the definition of statistical entropy at equilibrium. It has fundamentally altered this concept, which appears to be linked to the disorder present in an isolated system. As soon as a real system exchanges energy and/or matter, this concept of statistical entropy must be generalized since the state of the system under consideration is modified by various interactions. It can then evolve to a stationary state conditioned by maximum entropy production. Moreover, its generalization to quantum phenomena, that we have already introduced, sheds a new light on the properties of solids at the nanometric scale.

The concept of information, defined by Shannon in terms of probabilities, led to the extension of the thermodynamic domain to non-equilibrium systems. This is what we discussed at the end of Chapter 3, without considering all the consequences in computer science. The latter is indeed a technical declination of theories of information and communication [SEG 13]. Note that communication is defined in terms of transmission of information without analysis or validation of its content with information being the know-how, considered as a physical concept. When knowledge is acquired, the observer then becomes *de facto* integrated into the system in a subjective approach.

The experimental approach, with a key contribution by nanotechnologies that we will shortly introduce, will have fundamental consequences, in particular with the increase in computing power, which has become massive (Big Data). It results in the explosion of numerical simulations as a new tool for investigating an evolutionary system.

6

Nanomaterials

A practical extension of this theoretical recap is how to create nanoobjects and develop new nanotechnologies. We will first consider how these new solids are prepared and characterized by reviewing the main approaches, often oriented toward what is known as molecular electronics. The search for functional nanometric structures and devices began following Carter's proposals around 1980 [CAR 82]. One of the goals, as we saw in Chapter 5, is the development of computing, which predicts improved performances of the components due to their smaller dimensions and larger integration densities. This goes beyond Moore's empirical law and enters the realm of quantum physics over 10 nm [WAU 14].

For this purpose, we will review new synthetic pathways and their characterization techniques before examining their associated functions. Their reduction in size is accompanied by the search for new molecular components, in particular, issued from organic syntheses. We will briefly discuss these materials and the main molecular assembly and manipulation techniques, which have enabled their recent development.

6.1. The new classes of materials

There are two distinct methods used to manufacture these nanomaterials, where at least one of their measurements is at the nanometric scale. The first is that of an assembly of molecules crystallized in a one- or two-dimensional system (with a privileged plane or a symmetric axis) and with specific physical properties related to their spatial organization. The second approach is an induced organization at the molecular or macromolecular scale, which can have a size equal to or less than a nanometer. A finite set of molecules will exhibit a global behavior when subjected to an external stimulus; however, in the limiting case of a single molecule, there is no longer a possible statistical approach [WAU 14].

In this context, we will review the main groups of organic conductors for molecular electronics. This approach does not cover all new synthetic materials used, as we shall review other types of nanomaterials, but they do form the essential part [OFT 88]. Indeed, the fact that assemblies of organic molecules are conductors of electricity is probably one of the most significant advances in recent decades, even if their chemical stability can pose problems. These classes of compounds are based on molecular associations forming complexes by ionic interaction between donor and electron acceptor molecules. The possibility of electron transfer using a redox system and the stabilization of the complex by resonance energy forms the basis of the theory established by Mulliken and Person [MUL 69]. This phenomenon is closely related to the presence of π orbitals, resulting from the hybridization of atomic s and p orbitals for multiple carbon chemical bonds; it is associated with the notion of aromaticity and the presence of delocalized electrons.

6.1.1. *Conjugate conductive polymers*

Macromolecular chemistry and the synthesis of polymers were decisive steps during the 20th Century. The structures of several classes of conjugated polymers are shown in Figure 6.1, starting with the prototype series of polyacetylenes $(CH)_n$. These are electroactive polymers as they are subjected to a redox process; their π electron concentrations can change, which often leads to a change in color and the appearance of significant electrical conduction [HEE 01]. Note that, in this figure, the polyanilines are quite particular because the ionic conductivity is superimposed on electronic ones.

By controlling the charge transfer, plastic metals have been prepared in which the conduction process is particular, because they are one-dimensional electronic systems [OFT 88]. An advantage of this is the ability to produce molecular conductive wires that also present an electroluminescent effect, which we will develop further.

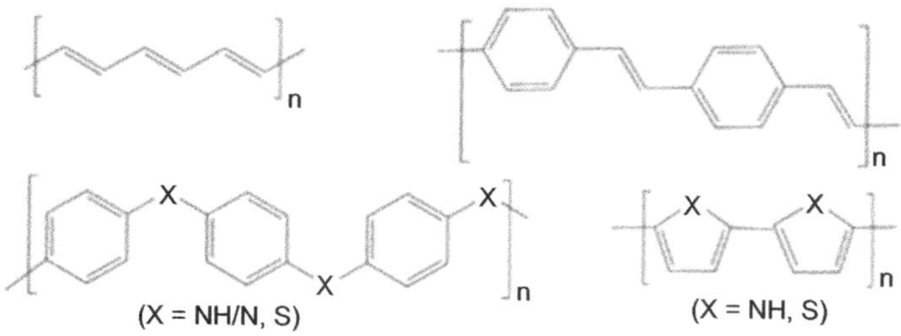

Figure 6.1. *Chemical structures of conjugated polymers (n is the degree of polymerization) for different families: (top left to right) polyactylenes then polyphenylene vinylene, (bottom left to right) polypyrroles (X = NH) and polyphenylene sulphides (X = S), finally polyanilines (X = NH/N) and polythiophenes (X = S)*

6.1.2. *Charge transfer salts and complexes*

These molecular complexes, defined by Mulliken and Pearson [MUL 69], may be in a neutral or ionic state if the charge transfer takes place spontaneously. In this context, we must distinguish between true complexes, the association of two radical ions, and salts where a radical ion is associated with a counterion without a single electron, just balancing the electrical charge: they are part of groups with mixed valence states. The first organic semiconductors were discovered in the 1950s [CAR 82] with salts of small aromatic molecules or tetracyanoquinodimethane (TCNQ). Figure 6.2 shows the main series of electron-donating molecules, cation-radicals obtained after oxidation, such as tetrathiafulvalene (TTF) and acceptors or anion-radicals, consecutive to a reduction such as TCNQ [DEL 90].

The spatial arrangement of these planar molecules gives rise to molecular complexes formed by different types of stacks. A crystal structure oriented in a preferred direction is observed, hence with a large anisotropy of the physical properties in general and their electrical conductivity in particular. The typical example is the TTF–TCNQ complex, which was the first organic crystal to exhibit metallic behavior. Belonging to this TTF series are salts derived of radical cations, the first organic materials exhibiting a superconducting state at low temperature with transition temperatures between 1 and 12 K. These solids are defined as having a mixed valence state, corresponding to partially filled electron energy bands. The elementary excitations are characterized by strong interactions, inducing a specific behavior due to the formation of Cooper's pairs (see Table 5.1). Their low electronic dimensionality gives them a peculiar behavior, described in terms of a quantum liquid. A superconducting but also a magnetic fundamental state compete with the appearance of a molecular ferromagnetism, which we will introduce in Chapter 7. Despite shaping problems, they were the first examples of organic electronic components, i.e. diodes and transistors [ISH 98].

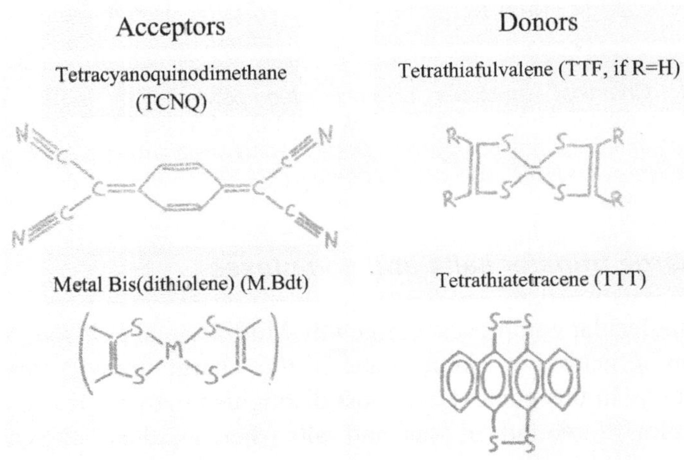

Figure 6.2. *Presentation and classification of the main chemical compounds forming crystalline molecular conductors (adapted from [DEL 90])*

6.1.3. *Molecular carbonaceous phases*

The polymorphism of carbon, based on the main existence of diamond and graphite, but also non-crystalline carbonaceous phases, is a consequence of the different possible hybridizations [LOI 06]. This topic has been completely renewed by recent discoveries of different molecular phases. In addition to a carbyne phase, which consists of an unstable linear chain of two-coordinate carbon atoms, they are formed from an atomic plane of three-coordinate carbon atoms mainly presenting hexagonal paving. The different varieties are shown in Figure 6.3, starting with the elementary brick of an atomic plane known as graphene. This chemical entity, which was chronologically the last to be isolated (Geim and Novoselov, Nobel Prize in Physics, 2010), is a polyaromatic entity where the π electrons are delocalized over the entire surface and provide good electrical conduction. These aromatic planes are stabilized by Van der Waals interactions, that is to say, an orderly stacking effect giving rise to the thermodynamically stable phase of graphite, which has a hexagonal structure. Paving with pentagons or hexagons results in topological distortions and the initial molecular block can be manipulated to give rise to fullerenes (see Figure 1.5) and nanotubes by a curvature effect with the closure of edges, which cost energy [GEI 07]. These molecular phases of concave shapes also tend to be found in multilayers, forming either graphene bulbs or multilayered onions or nanotubes. However, chemical synthesis has made it possible to synthesize and isolate defined molecular entities from single-walled nanotubes of controlled sizes and isomorphisms, but also isolated graphene ribbons (see Figure 6.3).

It is therefore semiconductors or one- or two-dimensional metals that have a specific transport mode called ballistic regime for charge carriers associated with optimum elastic properties. We will see that these molecules have led to significant advances in quantum molecular electronics.

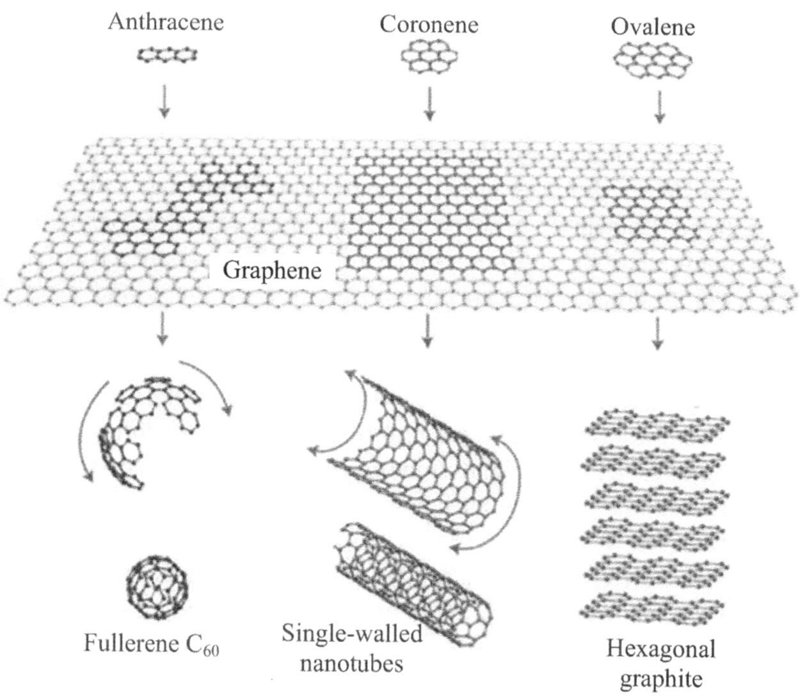

Figure 6.3. *Presentation of the molecular forms of graphitic carbons from condensed polyaromatic molecules serving as precursors for graphene ribbons and molecular varieties obtained after atomic plane curvature giving fullerenes and nanotubes (adapted from [GEI 07])*

6.1.4. *Other nanomaterials*

Numerous atomic or molecular assemblies of mineral or organometallic origin have also been proposed in connection with their surface chemistry [WAU 14]. Let us cite a few examples based on their shape symmetry, beginning with metallic gold wires and indium or vanadium oxides, to ensure an electrical connection. Next, heteronanotubes are obtained from isomorphic lamellar compounds of graphite. It is necessary to mention the electrical insulating nanotubes of boron nitride or molybdenite (MoS_2) [LOI 06]. Spherical aggregates have also been synthesized, such as, for example, silicon aggregates known as clathrates, metastable for certain stoichiometries (see Figure 1.5). Nanoballs of a controlled diameter are also prepared with metals or polymers or even hybrids, with a composition known as an organic–inorganic skin core. In general, these syntheses are carried out, starting from the atomic or molecular scale, according to a "bottom-up" organization.

6.2. Nanometric assemblies and manipulations

Advances in nanotechnology are fundamental in this field of nanosciences. We will not review the many innovations but instead analyze the main trends that have allowed the realization of molecular assemblies and their manipulations on the nanometric scale. This approach is based on two essential findings, the respective influences of shape and size of the objects created. As we saw at the end of the first chapter, the shape must be associated with a technical necessity; it is a crucial criterion that must be controlled by the experimentalist. Moreover, the role of surfaces or interfaces becomes significant: the limiting case is that with single-walled carbon nanotubes and graphene ribbons (Figure 6.3), where there are no more volumic atoms but only surface atoms. This must be functionalized to control exchanges with the environment, as we introduced in Chapter 3. In order to do this, different approaches aim to guide the construction of atomic or molecular structures to the fabrication of nanostructures with controlled shapes and special structures with controlled porosities [ROT 16]. This is achieved in two ways that organize the space and their interfaces: either through an external surface acting as a steric constraint to control an organization, or via mechanisms of intrinsic self-assembly. In particular, lithographic techniques are typical examples associated with the development of the silicon industry.

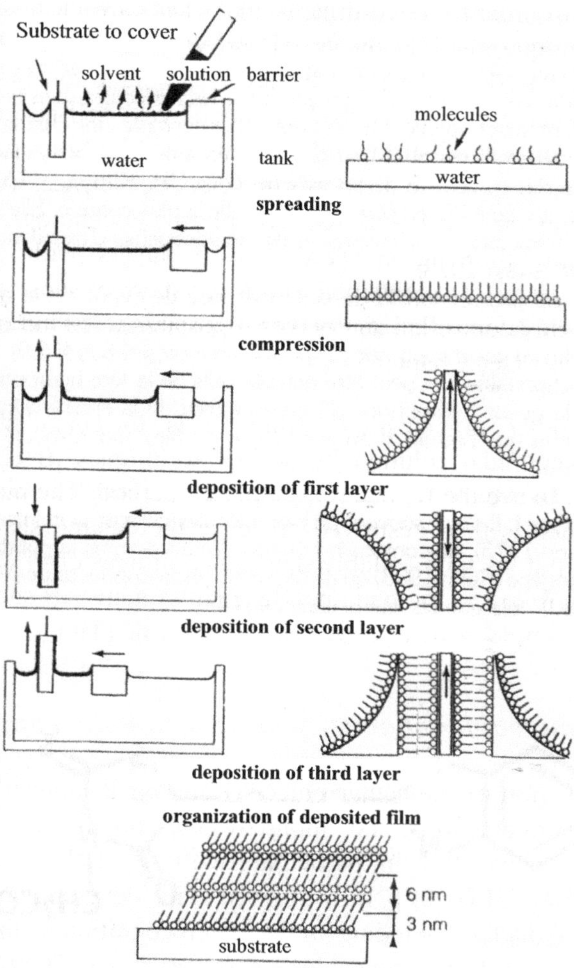

Figure 6.4. *Main steps of the Langmuir–Blodgett method allowing the successive deposition of single molecular layers (according to [OFT 88])*

6.2.1. *Thin film techniques and imposed structures*

The best known technique is the Langmuir–Blodgett technique, which requires the use of amphiphilic molecules, made of a hydrophobic part and a hydrophilic part [ULM 91]. As shown in Figure 6.4, at the air–water interface, a two-dimensional arrangement of such an insoluble molecule is obtained. This molecular layer is then transferred onto a substrate and different types of organized two-dimensional stacks are successively formed. A derived technique involves dispersed and water-oriented ionic polymers, which are transferred by the layer by layer technique to obtain a lamellar structure. Another approach is that of monolayers self-adsorbed by non-covalent bonds on a solid surface of elongated amphiphilic molecules. This approach makes it possible to perform selective surface treatments. For inorganic compounds, different reactive plasma technologies or vacuum molecular beam, allowing epitaxial growth, layers with a pre-defined thickness and shape are obtained on a crystalline substrate. This last technique is highly controlled and makes it possible to obtain quantum wells and metallic wires (see Chapter 7).

Finally, another organization technique is to coagulate nanometric particles of a colloidal solution by sol-gel techniques. It is applied in a selected matrix to use the replica (or template) method. The matrix serves as a preform, which is then destroyed, to leave the desired shaped structure, which is often porous with suitable shapes and dimensions [WAU 14]. This is the approach followed to obtain porous cellular materials such as foams or aerogels, for example made of active carbon or rutile (TiO_2).

6.2.2. *Supramolecular chemistry and the colloidal approach*

The construction of molecular objects from small molecules with weak intermolecular interactions, i.e. non-covalent bonds, is the basis of supramolecular chemistry. In particular, work by Lehn, Nobel prize in chemistry 1987 [LEH 09] has allowed the development of a self-organization of matter, including molecular recognition with the key-lock concept and in cage-shaped molecules such as cryptates. They may also have a specific catalytic character, also found in enzymes [OFT 88]. Molecular assembly, based on these weak and labile interactions, is present, for example, in liquid crystals by inducing a spontaneous orientation order. It is also a pathway to host–receptor specific biochemical assemblies with mesoscopic spatial organization inducing specific properties [SAO 13]. An example of a supramolecular scaffold using directional hydrogen bonds is shown in Figure 6.5. A monomolecular network of guanosines functionalized with oligothiophenes spontaneously orient themselves on a flat surface, as shown by near-field microscopy images (scanning tunneling microscopy [STM] technique, see Figure 6.9).

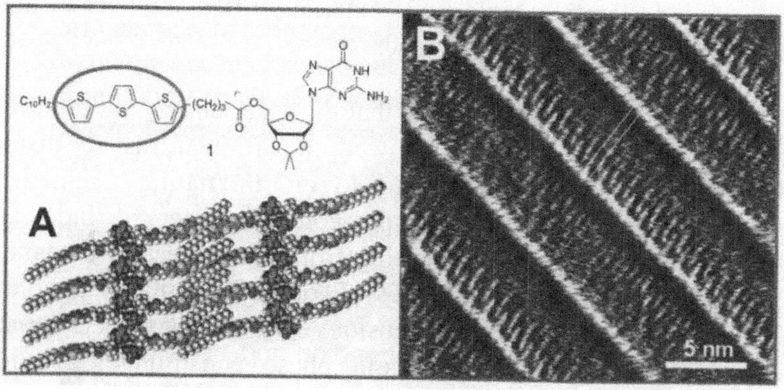

Figure 6.5. *Presentation of an alignment of guanosine-oligothiophenes with weak interactions between the thiophene and guanosime groups with the role played by hydrogen bonds: (a) the molecule and its schematic organization and (b) the topography of this two-dimensional organization observed using STM (according to [SAO 13])*

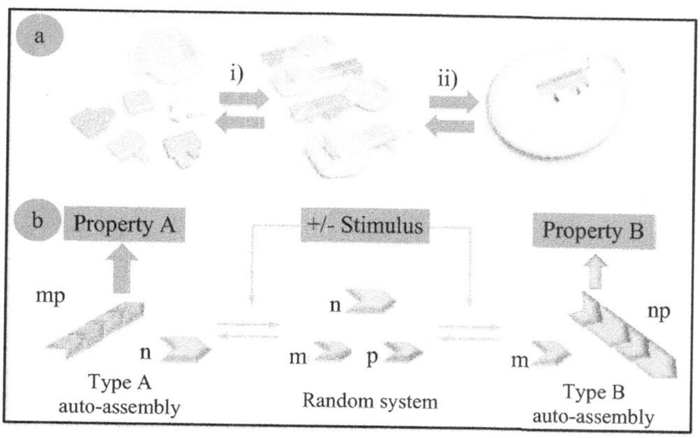

Figure 6.6. *Principle of dynamic combinatorial chemistry: (a) generation of molecular elements forming the library; (b) responses of a system to a specific external stimulus and leading to the formation of a type A or B self-assembly with different properties (according to [MOU 15])*

Dynamic combinatorial chemistry is the next step in building complex self-organized assemblies [MOU 15]. The principle of its implementation is shown in Figure 6.6(b), where the presence of an external stimulus modifies the type of assembly and provides information to the system. This adaptive and dynamic chemistry is an irreversible phenomenon that produces heat: it is a step toward adaptive biological materials, often multiscale, and then toward self-generated living environments.

Self-assembly in dynamic systems that dissipate energy started following work by Whitesides [GRZ 01]. He showed, in particular, that for dispersions of magnetic microparticles subjected to an alternating magnetic field, different types of mesoscopic aggregates could be observed. Indeed, in the so-called "bottom up" approach, starting from the molecular or nanometric scale, we must use the surface energies of colloidal chemistry to control the scaffolding. We recall that a suspension of small nanometric particles, dispersed in a continuous, liquid or solid phase, constitutes a colloid. This entity is in a metastable thermodynamic state where different phenomena of aging occur and surfactant molecules are generally used to stabilize them. Several synthetic routes exist, especially in liquid phase, to control their morphology and size. This is the case for gold nanoparticles [SEO 10] for which their shape, size and surface functionality can be controlled by chemical adsorption: a simple example is given in Figure 6.7.

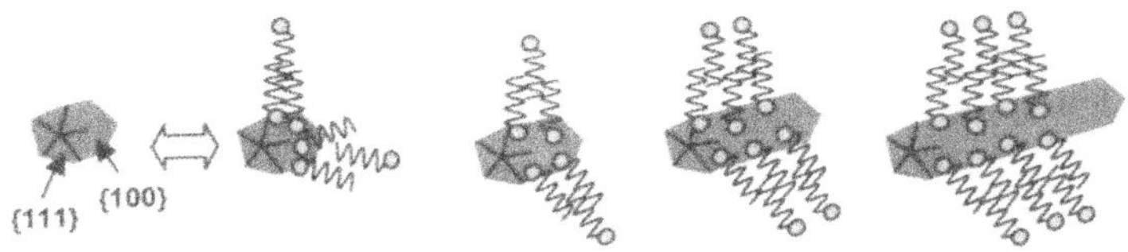

Figure 6.7. *Example of preparation of small gold rods by encapsulation in a layer of a surfactant from a polyhedral nucleus characterized by two different crystalline faces (according to [SEO 10])*

The assembly of small supramolecular aggregates by confinement in a drop can give geometric aggregates, leading to polyhedral associations but also colloidal crystals with different morphologies [PER 10]. Depending on the constraint exerted, one finds the fundamental polyhedra, already discussed in Chapter 1, or various stacks of spheres of a greater dimension. Indeed, it is possible to design multiscale entities with a size hierarchy and a complexity of forms leading to three-dimensional microarchitectures [SAL 17]. These materials can be hybridized with an inorganic core and an organic surface functionalized by chemical or electrochemical grafting, and even with the realization of porous or nanostructured surface materials. As we have seen for electrochemical storage in Chapter 4, the exchange surfaces are optimized.

6.2.3. *Nanowires and nanocomposites*

The distribution of nanoparticles in a solid matrix is another aspect of colloidal chemistry. These formulations lead to heterogeneous materials with a structural or a functional character. An example is that of the dispersion of single-walled nanotubes in an aqueous solution stabilized by a surfactant; the coagulation spinning process in a polyvinyl alcohol solution makes it possible to obtain a wire where the carbon nanotubes are initially oriented by flow [MOI 09]. The dispersion–coagulation device is shown on the left in Figure 6.8, with a photograph of the continuous spinning process on the right, making it possible to obtain a practical micrometric fiber.

Thus, nanocomposites can be considered as variants of conventional composite materials, which range from 1 nm in radial size to about 100 units in an organic matrix. The most common examples are multilayer nanotubes and carbon nanofilaments dispersed in a polymer. Charged polymers of these conductive nanoparticles, which are transparent, may be mechanically resistant and conduct electricity. These large variations are related to electrical conduction phenomena at the percolation threshold (see Box 8.2 in Chapter 8). They are associated with piezoconductive effects and conduction jumps under mechanical stress, characteristics allowing them to serve as electromechanical probes.

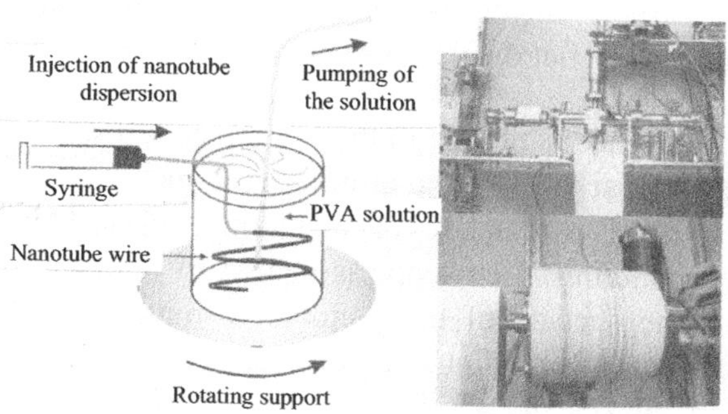

Figure 6.8. *Injection-spinning and rotation of a wire consisting of oriented single-walled nanotubes; on the left, a diagram of the experimental dispersion–coagulation device and on the right, a photograph of the prototype made for continuous spinning (adapted from [MOI 09])*

Other fields of application for nanocomposites exist, especially in cosmetics and health. For example, the dispersion of titanium oxide nanoparticles in a sun protection cream for skin or bactericidal silver nanoparticles incorporated in textiles. This problem is similar to that of modified coatings, inspired by plants (for example the surface of the lotus flower) and controlling the surface characteristics, such as the hydrophilic–hydrophobic equilibrium, their roughness and their wettability.

6.2.4. *Detection and manipulation of particles*

We will simply discuss two essential points in relation to the miniaturization of integrated and optical circuits. These are the observation and manipulation of objects due to the advances in new microscopies and optical systems.

6.2.4.1. *New microscopy techniques*

Associated with these advances, observation techniques at different scales have progressed considerably with the development of several electronic microscopies (see Chapter 1). From scanning microscopy in the mesoscopic range, there are various transmission microscopy techniques operating up to very high resolutions. They have allowed us to go down to the subnanometric scale and view atoms or molecules under certain conditions, beyond those of typical geometrical optics. In this context, it is necessary to point out the field emission phenomenon produced by the photoelectric effect. It is related to a tip effect that concentrates the electric field, present in filamentary carbon materials used as electron guns, for electron microscopy and also for flat display screens.

A significant advance is related to the more recent discovery of near-field microscopies, particularly STM and atomic force microscopy (AFM), which allow the analysis of conductive or insulating surfaces [RAB 89]. The principle of the tunneling microscope is shown in Figure 6.9, where a live metal tip skims the surface of the conductive material and an electronic current is observed. By scanning the surface, a variation of this current is recorded, which makes it possible to reconstitute the relief and the topography of this surface, in particular its roughness and its nanostructure. This device makes it possible to follow the mechanism of depositing atoms or molecules on a flat surface and to manipulate them to produce a pattern. It is thus possible to manufacture objects on the atomic scale and generate particular atomic or molecular assemblies. To illustrate this technique, an example of an STM image has already been presented in Figure 6.5 [SAO 13].

Finally, in this context, optical fluorescence microscopes functioning with specific markers (e.g. rhodamine or pentacene) must be mentioned. From a stimulated emission of the characteristics below, the classical diffraction limit resolution can be exceeded up to the observation of a single molecule [MOE 12]. These nanophotonic techniques, developed in particular by Moerner (Nobel Prize in Chemistry 2014), enable local observations in various biological media.

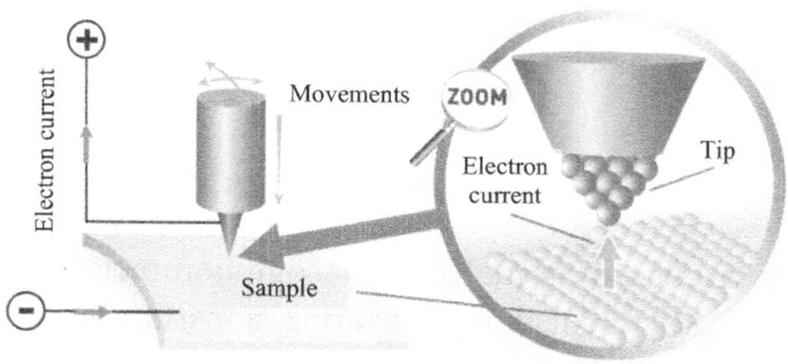

Figure 6.9. *Principle of operation of an AFM near-field microscope by tunneling an electronic current detected on a surface sample*

6.2.4.2. *Optical clamps*

These are unique tools because they allow the manipulation of an object without any physical contact, unlike mechanical nanotweezers. In fact, optical clamps exploit the radiation pressure exerted on any surface exposed to electromagnetic radiation [COI 03]. If a bead is placed in a gradient of light intensity because of a laser, it has a stable focal point. In the vicinity of this point, there is a shrinkage called a neck, where more intense light traps the particle, characterized by a retraction force F (see Figure 6.10). The displacement of the laser beam in an open medium induces the manipulation of particles with size of the wavelength of the light beam. It makes it possible to measure nanometric displacements and the forces exerted (of several picoNewtons only). An example has been already given (see figure 5.1) and numerous applications in cellular biology have emerged [COI 03].

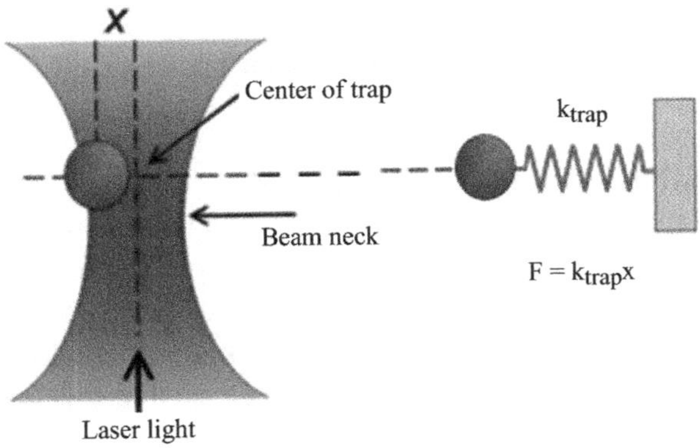

Figure 6.10. *Operation of an optical clamp: on the left, trapping of a particle in the neck of a laser beam, and on the right, a model of the associated elastic force (F) (where k trap is the constant proportional to the displacement x)*

6.2.5. *Molecular recognition, nanosensors and actuators*

Nanosensors are based on molecular recognition mechanisms, i.e. shape and functionality described above, ensuring a particular physicochemical characteristic. For example, they allow the environment to be controlled or to activate a specific function in the case of a clinical or biological diagnosis. As for conventional sensors, there are two stages: the recognition of a chemical signal and its conversion into an electrical or optical signal. The receptor is based on the key-lock molecular recognition effect, which allows high selectivity [HU 14]. The transducer transforms detection into an electrical signal in an integrated device that is sensitive to selective trapping. Typically, this detection, which must be sensitive and rapid, will result in a variation in electrical conduction or a response by a field effect transistor or chemical field effect transistor. The smaller size compared to conventional sensors has allowed new applications such as the detection of harmful gases.

Moreover, the transformation of an electrochemical signal into a mechanical actuator in nano-electromechanical systems is a miniaturization of micro-electromagnetic devices (MEMs) presented in the Chapter 3. These devices can serve as a detector of specific molecules using a non-covalent interaction (for example helical winding of DNA a strand or other biopolymers around a carbon nanotube) or as a source of mechanical force, created from of an electrochemical cell. This is the case for carbon nanotube wires proposed as an actuator for creating an artificial muscle; this is an example of the conversion of electrochemical energy into mechanical energy [LI 08].

Another class of deformation actuators for programmable and controlled machines, up to flexible robots, has been developed due to new 3D printing techniques [MOR 14]. Structures formed of elastomeric polymers can be assembled by creating Lego bricks (see Figures 6.11(a) and (d)). These layers are stacked and clicked together (Figures 6.11(b) and (f)) recalling the above-mentioned layer-by-layer technique. Polymerization of the elastomer may occur and result in a particular hollow shape (Figures 6.11(c) and (g)). Finally, under pneumatic pressure, an anisotropic deformation is obtained, which depends on the stacking of the initial layers. Thus, a convex cylinder and localized swelling can be obtained (final shapes shown, respectively, on the right and left of Figure 6.11). It is an example of complete control of an object's final shape according to its purpose.

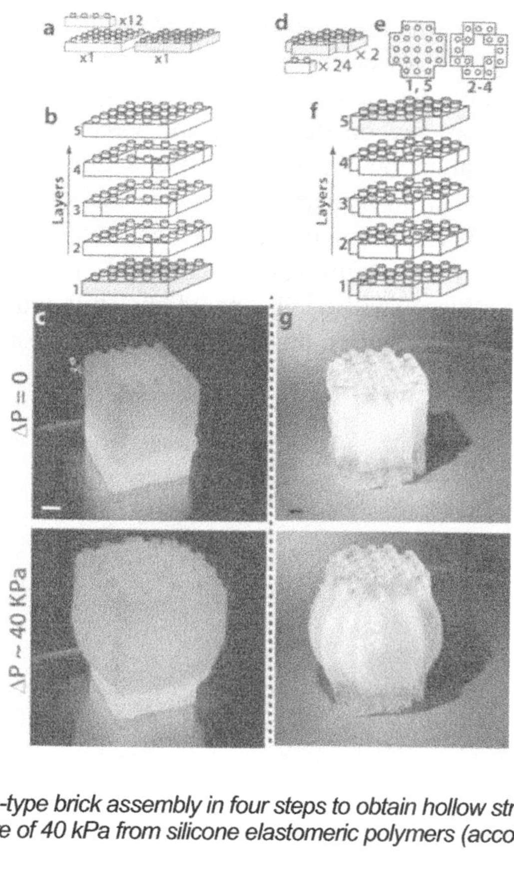

Figure 6.11. *Lego-type brick assembly in four steps to obtain hollow structures formed by an internal pressure of 40 kPa from silicone elastomeric polymers (according to [MOR 14])*

6.3. Conclusion

The aim of this transitional chapter was to present new synthetic components, especially those created in organic or organometallic chemistry; they are often used as advanced or adaptive nanomaterials. In particular, the discoveries with regard to these molecular materials of cooperative quantum properties, such as superconductivity and ferromagnetism, support the thermodynamic classification of materials. It is no longer the chemical origin of a solid that allows the classification of its properties, and a more comprehensive approach becomes necessary.

In this context, the main topics of these new nanotechnologies, relating to their preparation and their characterization, were reviewed as they are fundamental in the current state of affairs. Indeed, the discovery of new observation techniques makes it possible to cover all size scales down to the atomic scale. As we saw in Chapter 4, elementary electronic components are becoming smaller and smaller in order to increase the execution speed and information density at a lower energy cost. One must be able to print ever smaller circuits using improved lithography techniques. This fabrication of nanostructures involves several steps, with the spreading of a resin and its etching by UV radiation on the surface of a single silicon crystal. The practical problem is then to manage the heat evacuation produced by these logic systems and microprocessors (see the Landauer principle and Figure 5.2). Finally, the most significant theoretical consequence of this miniaturization of objects will be the paradigm shift introduced with the efficiency of quantum mechanics: this is what we will investigate further.

7

Engineering and Molecular Electronics

In this chapter we will highlight the main areas where significant advances have been made using nanomaterials and associated technologies, which we introduced in Chapter 6. These phenomena at the nanometer scale reveal their quantum nature based on the foundations described in Chapter 5. We will discuss the functional classification mentioned in Chapter 3, considering elementary excitations such as electrons, with associated charge and/or spin, and also photons as well as the electron–photon interactions. Thus, we will specifically present nanoelectronic systems, followed by developments in nanophotonics and then nanomagnetism. The concept of a molecular motor, which describes a supramolecular assembly capable of performing mechanical movements, will then be presented. These will be grouped under the term nanomachines.

With regard to various experimental devices, the problem of processing quantum information arises, i.e. the storage and reading of memories relating to individual particles or grouped in small colonies. The associated logic in this new engineering will be analyzed in the second part. It should be noted that this review is not exhaustive, but aims to show the most significant devices and tools in this developing field.

7.1. Nanotechnologies

7.1.1. Nanoelectronics

This is the most promising sector, with two types of approach, either at the ultimate scale of the isolated molecule or associated with a more mesoscopic dimension concerning an assembly of molecules. Before stating the main domains, note that in organic electronics, the delocalized π electrons play the role of conduction electrons obeying the Fermi–Dirac statistic (equation [5.5]). In conductive molecules or macromolecules, the conduction regime is no longer classical, i.e. diffusive, but occurs often in a ballistic regime, as introduced in Chapter 6. Indeed, an electron can pass through a sample of a reduced size without undergoing inelastic collisions and behave like a massless particle: it then presents a wave-like character. This is the case in particular for single-walled carbon nanotubes and graphene ribbons (see Figure 6.3), which are electronic low-dimensional systems [DEL 11a].

7.1.1.1. *Transistors and logic circuits*

After many attempts to produce field-effect transistors, based on organic molecules such as polythiophene, the most convincing results are actually obtained with molecular carbons. Single-wavelength semiconductor nanotubes have enabled the construction of field-effect molecular transistors with interesting characteristics relating to silicon technology (see Figure 3.4). In particular, a significant amplification factor with a high response frequency has been observed [PEN 14]. This quantum approach developed over several years has been replaced by involving graphene ribbons. The preparation of a conductive and transparent graphene sheet by epitaxial deposition on a silicon carbide (SiC) substrate is a nanolithographic technique, which also makes it possible to create a molecular transistor [WAS 10]. In Figure 7.1, such a device is schematically represented with a nanoscale grid at the center controlling the circulation of the charge carriers between the source and the collector in the doped SiC layer and therefore behaving as an extrinsic semiconductor. Miniaturization must lead to an optimal density of components operating in radio frequency and with minimum dissipated energy. The design of these components leads to the single-electron transistor. Indeed, in the substrate under electrical tension between the source and the collector, an electron pocket is formed and when a critical number is reached, the current no longer passes. This phenomenon is known as Coulomb blockade (see section 5.3.3 in Chapter 5). By modifying the gate voltage, the electrons can be entered or blocked one by one.

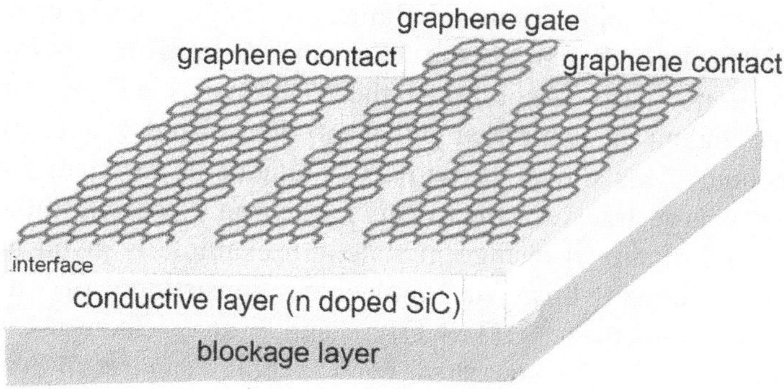

Figure 7.1. *Model of a field-effect transistor with graphene ribbons deposited on a silicon carbide support and operating in the quantum domain (adapted from [WAS 10])*

Thus, high-performing nanotransistors have been produced, but technological problems still remain. In fact, it is necessary to manufacture them reproducibly by classifying molecular objects from their shapes and physical characteristics. Then, it is necessary to manipulate and organize them to design and carry out reliable logical operations, as in classical microelectronics. An adapted architecture redefining logical circuits must be developed industrially: this is the current challenge.

7.1.1.2. *Examples of molecular compound*

In these devices, all the elements of a circuit including the connection must be on the same scale. Various functions have been described, starting with the conductive molecular wire and then also molecular switches. The most cited macromolecule are doped polyacetylenes or parent polymers (see Figure 6.1), capable of transmitting information via the transfer of energy or electrical charges. A complementary approach involves intramolecular diodes and switches. A convincing example is a single molecule junction between polythiophene and a chromophore, which exhibits electroluminescence and behaves as a molecular light-emitting diode (LED) [CHO 16].

Several types of donor or acceptor molecules have been identified from the initial proposals of Aviram and Ratner to design a molecular rectifier [AVI 74]. They are based on the following formulations between two acceptor or donor parts (A or D) of the electrons: A-π-D or A-σ-D; the intermediate ligand is composed either of delocalized π electrons or a localized σ bond. This allows them to act as a diode or a switch: the twisted intervalency charge transfer effect occurs when an electronic modification leads to a conformational change in state. An example is given in Figure. 7.2, which shows how a molecule consisting of a donor (tetracyanoquinodimethane (TCNQ)) and an electron acceptor (tetrathiofulvalene (TTF)) separated by a σ bond can result in a diode between two metal contacts.

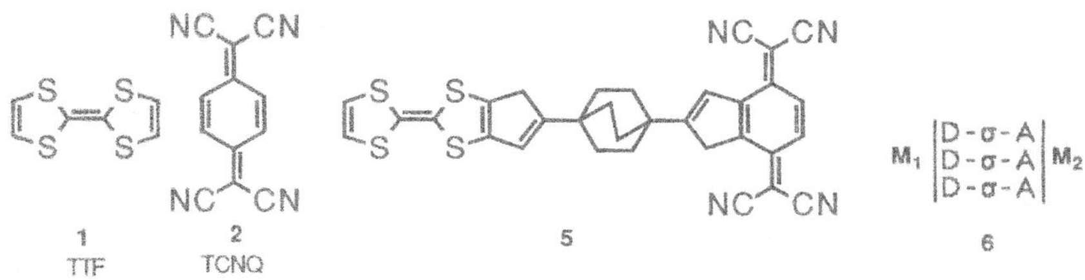

Figure 7.2. *Principle of an organic D-σ-A diode (formula 5) located between two electrical contacts M_1 and M_2 to control a charge transfer in (formula 6) one direction only (adapted from [AVI 74])*

Finally, a similar concept is that of molecular bistability, or in other words, how a single molecule can have two neighboring energy states switched by conformational change under the effect of light and thus controllable by this external stimulus. This is the case, for example, of organic photochromic molecules [AVI 74]. Another example is that of a charge transfer complex, which can pass from a neutral to an ionic state by phototransfer and exhibit hysteresis during the transition. This concept of spatial bistability has already been introduced for a non-equilibrium macroscopic thermodynamic system (see Figure 2.4). It is also extended to a finite set of molecules defining a microscopic system subjected to a stimulus. The transition from a statistical unit whose components can interact to the limiting behavior of a single and isolated molecule, however, appears to be a topic requiring further investigation.

Ultimately, this approach is defined for a single molecule that can store information if it has two nearby coupled energy states. This situation corresponds to a model with two potential wells that an external perturbation can selectively address [KAH 88]. Such an attractive approach still poses conceptual problems and with regard to addressing, connecting and reading techniques for a single molecule or a small defined set. Progress remains to be made in this context; the realization of sets organized due to a particular molecular film assembly by Langmuir–Blodgett technique has led only to positive results in the laboratory.

7.1.2. *Nanophotonics*

The laws of geometrical optics are modified and surface phenomena become preponderant. A simple nanometric hole on a metallic surface induces a particular optical transmission [HE 09]. The presence of surface plasmons, as elementary excitations due to collective oscillations of electrons on a surface, plays an essential role. Semiconductive lamellar solids and graphene are the preferred materials. In general, quantum effects, with increased and controlled emissions of photons, are observed. We will list the various effects associated with the transmission or emission of a light wave.

7.1.2.1. *Photonic crystals and metamaterials*

Colloidal photonic crystals are periodic dielectric nanostructures with a pattern in the range of the used wavelength in one, two or three directions of space. They are formed using various methods of colloidal chemistry such as, for example, the Langmuir–Blodgett technique enabling the controlled deposition of a stack of molecular layers of an amphiphilic entity. One proposed application is the manufacture of artificial opals by regularly stacking micrometric silica bead layers.

Laser generation and light transport can channel and store photons. It is possible to produce a silicon waveguide and to store optical information in a resonator, a Fabry–Perot type resonant cavity (see Figure 3.8). The production of an optical transistor for coding binary language information and producing logic gates, in particular with organic molecules, is an initial approach toward an optical computer [CLA 10].

Another class of heterogeneous medium is that of polyphasic materials, when their characteristic dimensions are less than the optical scanning wavelength. They can then have a negative refractive index, a property unknown in nature. They are optical or electromagnetic metamaterials, depending on the size of the entities; their analogues with phonons are acoustic metamaterials. For the former, such as a meta-surface of nanometric holes, it is possible to obtain a high-performance nanoantenna, an interesting characteristic for light, and act as an optical lens with increased resolution [KHO 16]. In this sector, potential applications, linked to the creation of organized systems, are currently under development, as demonstrated by the creation of a layer deflecting the visible light. An object covered with a metamaterial layer becomes invisible for a given wavelength [NI 15]. These results are relevant of the more general frame called nanoelectromagnetism, defines as the interaction of electromagnetic waves with nanostructures.

7.1.2.2. *Fluorescent nanocrystals*

This is a feature of the quantum box shown in Figure 5.3. The spatial confinement of a nanocrystal of a given size allows for the emission of monochromatic radiation. It is a fluorescence emission of visible light under illumination in the ultraviolet spectrum, as in the case of cadmium selenide or organic molecules. The only experimental parameter, allowing the emitted color to be chosen, is the size of crystals that are formed by means of colloidal chemistry. An example is given in Figure 7.3, where cadmium selenide nanocrystals, ranging from 2 to 6 nm, are stabilized by the adsorption of surfactants. Absorption decreases as sample size increases causing the displacement of the fluorescence peak [FER 08].

Two major applications include fluorescence markers in biophotonics and for display screens. By selectively fixing molecules, in particular biological molecules, on the treated surface of these nanocrystals, markers functioning in vitro or in vivo have been produced [HAN 15]. The second area of application, derived from the fluorescence microscope, is their use in screens to improve visual contrast and imaging techniques. For example, with a quantum dot device placed in between microtweezers, an emission of single and indiscernible photons could be achieved with good efficiency [DIN 16].

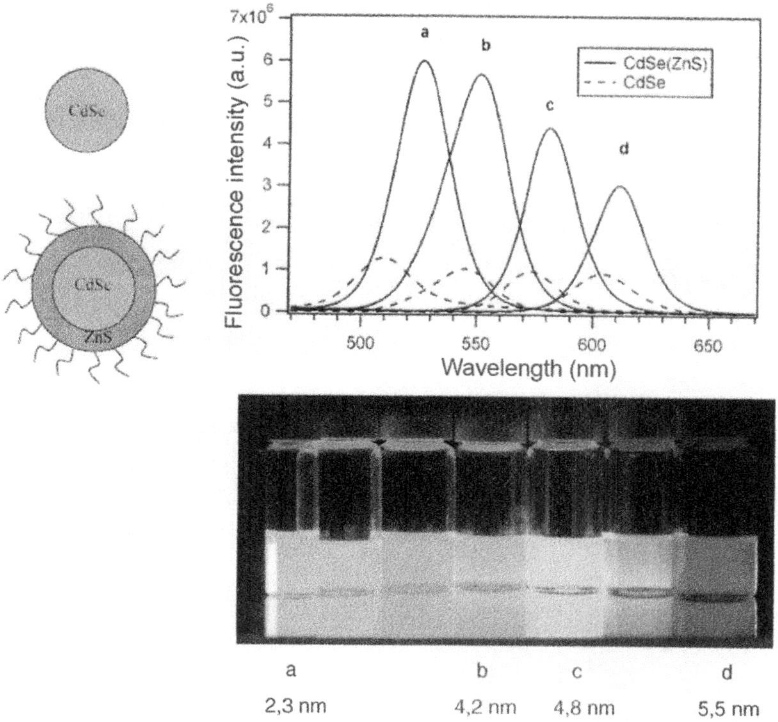

Figure 7.3. *Fluorescence emission of CdSe quantum dots, pure and doped (with ZnS), and stabilized by surfactants; seven samples of increasing size show the displacement of the color emitted from blue to red, relating to the variations in forbidden band width (adapted from [FER 08]). For a color version of the figure, see www.iste.co.uk/delhaes/materials.zip*

7.1.2.3. *Organic electroluminescent diodes*

The photoelectric effect is observed in diodes made with single-walled nanotubes that can emit light. Upon recombination of an electron and a hole at the p-n junction, and the formation of an exciton with the emission of a photon, this is the inverse process of the photovoltaic or LED effect already presented (see Figure 3.8). We have noticed this effect from conjugate polymers [CHO 16], but also doped semiconductors (see Figure 6.1), or even organometallic complexes of many devices emitting white or monochromatic light have been developed. Organic LEDs developed over the past 30 years [TAN 87] have undeniable advantages despite a somewhat limited lifetime. They make it possible to produce emissive screens and develop organic optoelectronics on adaptable flexible surfaces by simplifying the connectivity difficulties and reducing the energy dissipation by the Joule effect. A key example is that of a single-walled carbon nanotube photodiode presented in Figure 7.4 [MUE 10]. Indeed, at the p-n junction, an efficient mechanism for emitting light on a reduced spectrum has been shown, suggesting possible uses.

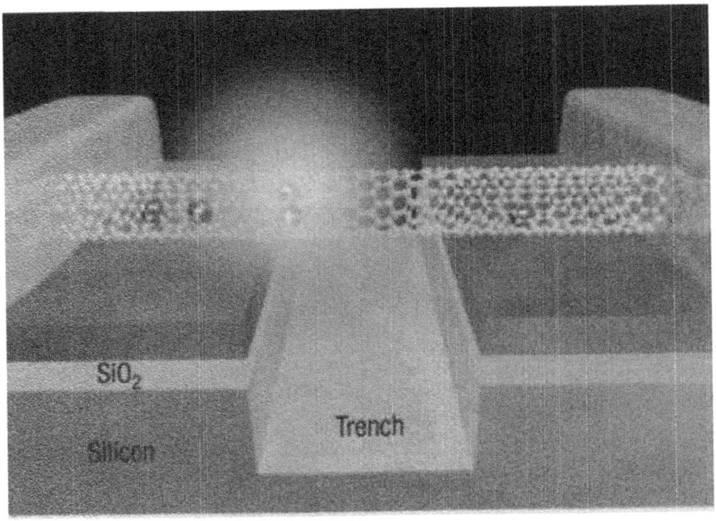

Figure 7.4. *Schematic diagram of a carbon nanotube photodiode with the emission of intense light concentrated in a wavelength range (adapted from [MUE 10]). For a color version of the figure, see www.iste.co.uk/delhaes/materials.zip*

Finally, the emission of visible light not by a photoelectric effect but by a thermal mechanism on a graphene layer on a substrate has been published. This is the return of old incandescent filament lamps but with an increased radiative emission [NAK 08].

7.1.2.4. *Photochrome molecules*

Under the action of light, a photochromic molecule is chemically transformed and exhibits a different UV-visible absorption spectrum. Several families of organic compounds exist, characterized by the switching speed associated with a conformational change by isomerization in the presence of frequency and light intensity constraints [KIM 15]. These reversible dynamics can also be associated with a photomechanical effect: three significant examples at different scales are presented in Figure 7.5. This phenomenon can be used as a photoswitch to create logic gates or, more recently, as a metal ion detector [HUA 15].

These molecules (e.g. spiranes or spiro-oxazines) are dispersed in a plastic or directly synthesized in a photochromic polymer. The applications concern the control of photon transmission in optical glasses or for intelligent windows. Finally, on a molecular scale, they are potentially bistable optical systems for storing information that can be associated, for example, with an elastic or magnetic behavior and provide elements for a future optical computer.

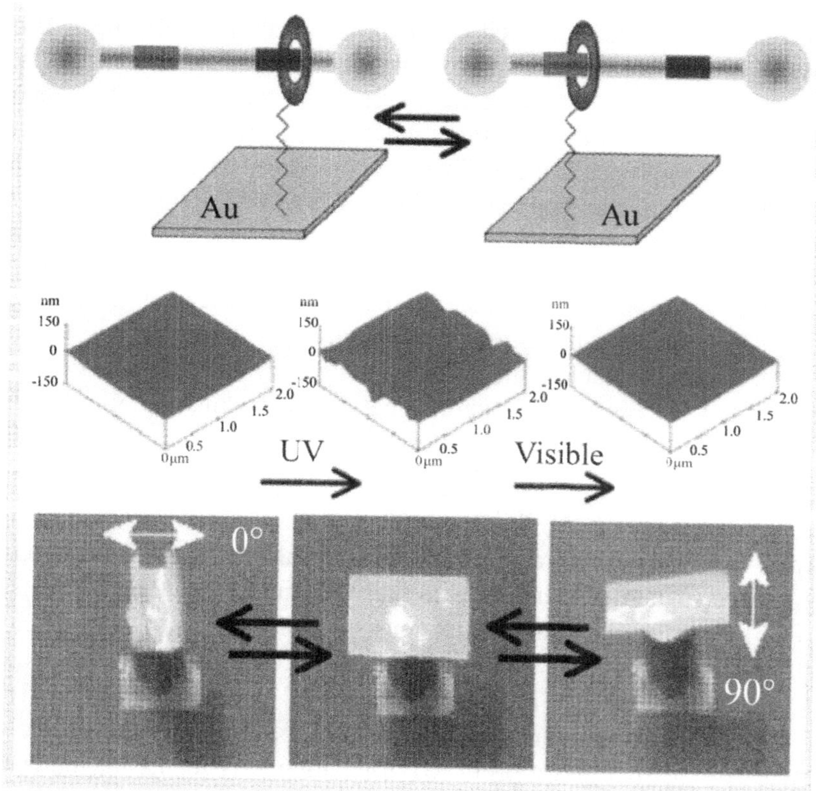

Figure 7.5. *Photomechanical effect: an example of molecular shuttle by photocommutation of a fluoro rotaxane adsorbed on a gold surface (top); photoinduced topographic changes in the UV and visible spectrum observed by nanometric AFM microscopy (middle); macroscopic and reversible photomechanics (black arrows) on a polymer film deformed by different polarizations of the light wave (white arrows) (adapted from [KIM 15]). For a color version of the figure, see www.iste.co.uk/delhaes/materials.zip*

7.1.3. *Nanomagnetism*

Note that electronic magnetism is based on the presence of a set of electronic spins (s = 1/2) and an associated magnetic moment. On the macroscopic scale, cooperative magnetic states exist, which were introduced in Chapter 3 and which have to be interpolated until the case of a single molecule. At the end of this chapter, we introduce the characteristics of a cooperative set of spins (see Figure 3.6 for the definition of a hysteresis loop). Following this way, we are going to focus on magnetism at the local level that has developed greatly in recent decades. Techniques for the fabrication of metallic nanoscale islands have made it possible to create single magnetic domains, which can exhibit bistability [DIN 05]. Another approach in the miniaturization of magnetic thin-film memories is the realization and control of magnetic Skyrmions, which are stable quantum quasi-particles such as those defined in Table 5.1 and described as swirl of matter on the topological aspect. They are considered as very small magnetic bits for future storage and processing devices [SAM 13].

By exploiting the flexibility of molecular chemistry, two other major axes have been developed with the discovery of molecular nanomagnets [KAH 93] and spin-transition compounds. In both cases, they are organometallic molecules in which magnetic bistability is present, supplemented by reversible photomagnetic effects, making it possible to envisage the molecular storage of information.

7.1.3.1. *Molecular nanomagnets*

Apart from stable free radicals and ion-radical salts based on fullerene or TCNQ, experimental evidence was initially obtained more than 20 years ago [GAT 96]. A molecular nanomagnet consisting of a stable polyatomic cluster comprising 12 manganese atoms (see Figure 7.6 showing the structure of $Mn_{12}\ O_{12}$ (carboxylate)$_{16}$). It has two degrees of oxidation Mn^{3+} (s = 2) and Mn^{4+} (s = 3/2) with an overall molecular spin of S = 10. This molecule with a mixed valence state has a preferred orientation axis for its magnetization. At a very low temperature (T = 2.1 K) the magnetization (M) measured along this axis shows a magnetic hysteresis loop. It is not due to the presence of magnetic domains that would align gradually, but to a quantum tunneling effect with the orientation of the atomic spins changing from the parallel to the antiparallel state. Between two potential wells separated by a magnetic anisotropy barrier, this quantum spin reversal phenomenon gives rise to steps corresponding to intermediate magnetic states.

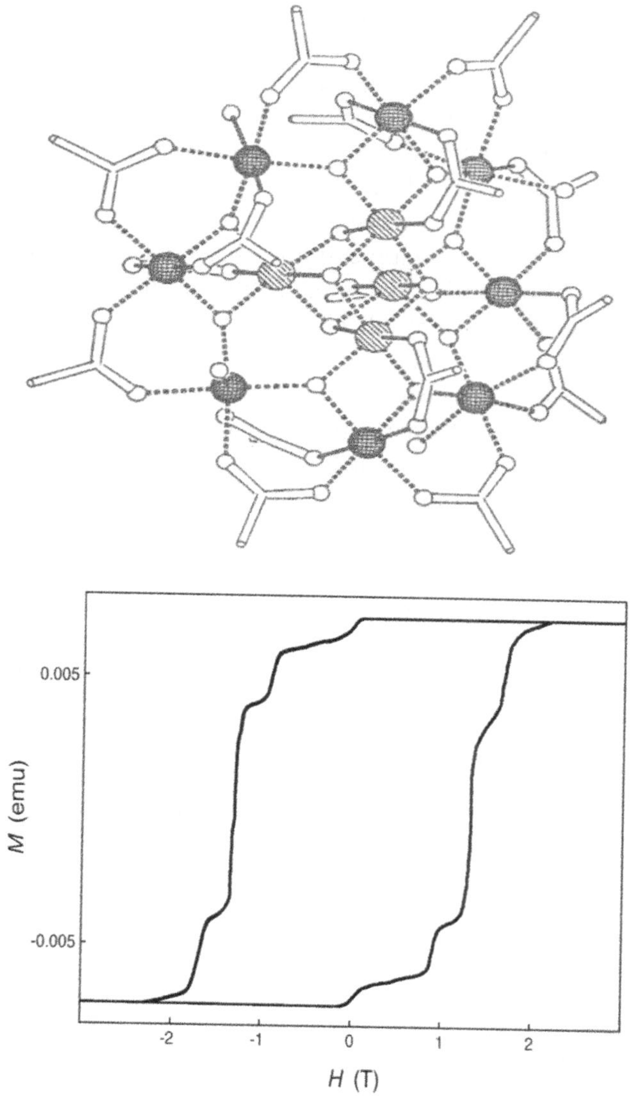

Figure 7.6. *Crystallographic structure of the molecule Mn_{12} O_{12} $(CH3COO)_{16}(H2O)_4$, the 8 Mn_{3+} ions are stripy and the 4 Mn_{4+} are hatched; beneath, the magnetization cycle M is observed at 2.1 K showing a slow variation with quantum steps under the external magnetic field H (adapted from [KAH 93])*

Other organometallic complexes exist such as chains of magnetic molecules that are one-dimensional magnetic systems, or the Prussian blue family consisting of an octahedral brick: $M^{3+}(CN)_6$. With M = Fe, Cr … and associated with another M'^{2+} ion. They are bimetallic compounds of transition metals (Fe, Cr ...) in two different valence states and exhibit a ferromagnetic coupling at low temperatures [GAT 96]. In an aggregate of atoms or molecules, it is a so-called superparamagnetic state that is observed below a blocking temperature. The interaction between the spins is related to the quantum phenomenon of exchange, described by a Heisenberg Hamiltonian that is of an intramolecular nature [VER 01]. This characteristic is demonstrated by the comparative studies between the crystallized state and the Langmuir–Blodgett layered organization or diluted in a magnetically inert matrix showing the same phenomenon [MIN 00]. These molecular magnets can undergo a magnetic transition under the influence of temperature or pressure but also by selective light irradiation. The aim is to store molecular information on rewritable photomagnetic supports. In this context, the combination of magnetic bistability with an optical bistability is a promising approach that leads to a tristable state [FEN 13].

7.1.3.2. *Spin transition molecules*

A paramagnetic transition metal ion in an octahedral environment may have a high spin state or a low spin state, with parallel or antiparallel spins. Depending on the relative values inside an organometallic complex, between a molecular parameter known as the ligand field and the spin pairing energy of the electrons concerned, both situations are possible. They are functions of the external supply depending on the external supply of energy in a closed system [GUT 96]. This is the case for complexes formed of different ligands with (Fe^{2+}) with a high spin transition state (S = 2) and low spin state (S = 0) transition, an example is shown in Figure 7.7.

Hysteresis is observed, accompanied by a structural phase transition, but can be removed by quenching (dark points for decreasing temperatures and triangles for heating). These hysteresis loops, due to a cooperative effect, are not always observed and progressive conversions exist obeying the classical Maxwell–Boltzmann statistics. Photomagnetic effects have been discovered, in particular the light-induced excited spin state transition effect or even because of a photoisomerization of the ligand, acting on the spin state of the metal. A light disturbance, causing a change in the spin state, can be used in display devices by changing color instead of temperature or pressure intensive parameters. Indeed, these cooperative phenomena must be controlled at the molecular level in order to store information. Thus, the concept of a single magnetic magnet opens the way to nanospintronics in the field of quantum computing [LUI 13].

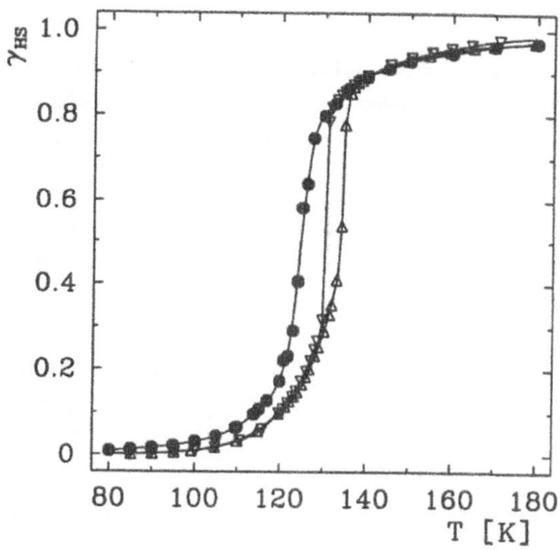

Figure 7.7. *Example of a thermal spin transition for a coordination compound Fe (ptz)(BF$_4$) where the high spin state fraction Y_{HS} is determined experimentally from a magnetization measure, which is plotted as a function of temperature (according to [GUT 96])*

7.1.4. *Nanomachines*

The energy required for a moving molecular machine can be of chemical, electrical or luminous origin. For these energetic conversions, the last two cases are electrochemical or photochemical processes: in fact, we have already given an example in Figure 7.5 with the presence of a shuttle on a molecular wire. Let us also recall that in Chapter 3 we already introduced materials far from equilibrium that react to an external stimulus. This is the case of electro-, photo- or chemosensitive materials that cause a mechanical effect such as contraction or reversible elongation [SAU 16]. These molecular machines are actuators such as nanoelectromagnetic (NEMs), which act under the influence of electrical energy in the case of carbon nanotubes, proposed as artificial muscles at the end of Chapter 6. Other supramolecular functions have also been introduced, essentially as a shuttle and a switch, or they can induce a translational or rotational movement (Sauvage, Stoddart and Feringa Nobel Prize in Chemistry, 2016). Two essential types of nanomachines have been realized:

– based on the molecule rotaxane (see Figure 7.5), which is shaped like a dumbbell and with a ring whose movements are controlled chemically [KIM 15]. These evolutionary self-assemblies also include interleaved molecules and pave the way for a topological approach [STO 09];

– molecular rotary motors make it possible to produce a nano car. A "circular" molecule has two rotor blades with a preferred rotation in a direction determined by light radiation or the application of an electric force. It constitutes an elementary wheel of a vehicle as represented in Figure 7.8: such an assembly rolls on an atomic roadway [RUD 11]. Other devices have been made, with it being possible to rotate the wheels in both directions due to the use of a near-field microscope [PER 13a].

These applications can extend to *in situ* therapies, which will be presented in Chapter 8.

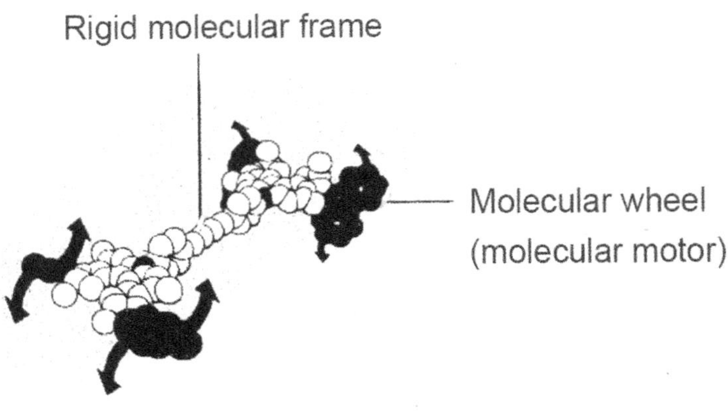

Figure 7.8. *Schematic structure of a molecular vehicle with four wheels propelled by a change in conformation under an external constraint (adapted from [RUD 11])*

7.2. Memory and quantum logic

We will describe the general concepts associated with quantum computing before analyzing the main experimental approaches to realize an operational quantum system. For this, the main advantages and disadvantages will be discussed. To conclude, a link with statistical thermodynamics in these quantum systems will be made.

7.2.1. *Quantum phenomena*

Quantum computing is based on two phenomena that do not exist in classical mechanics: superposition and entanglement. They define the properties of the qubit that we introduced in section 5.3.3 of Chapter 5. These two characteristics make it possible, on the one hand, to process the information more efficiently and, on the other hand, to transport it quicker: this is what is known as teleportation [BEN 13]. To summarize this situation, these quantum characteristics are presented in Figure 7.9 with the concept of superposition of quantum states and the phenomenon of entanglement between two objects that are described together only in order to be teleported. In this context, it is necessary to remember that quantum systems are fragile and affected by observations, which is not the case for classical macroscopic systems. This perturbation plays on the phenomenon of entanglement and the associated coherence time, which are modified by physical measurements; it takes a quantum technique to recopy information and transport it without destroying it.

These behaviors intervene at the particle level as an isolated atom and are based on the manipulation of individual quantum systems. The fundamental phenomenon of coherence–decoherence was studied by Wineland and Laroche (Nobel Prize in Physics of 2012). Thus, the coherence time of a quantum state must be as long as possible and protected from external perturbations [HAR 15]. It is then possible to address and transfer the information reproducibly over a great distance; this problem has recently been solved by covering kilometric distances using teleportation. For this purpose, it is necessary to manufacture the physical supports with qubits satisfying certain requirements. We will now summarize the main technical advances.

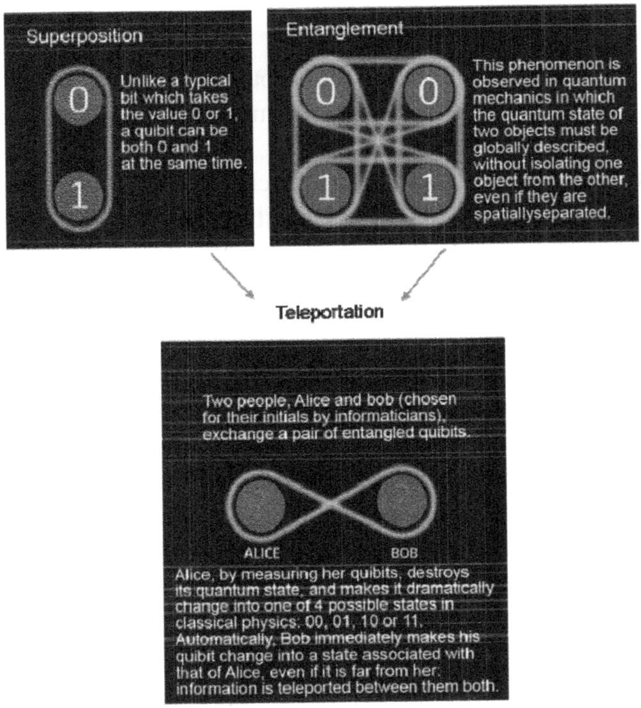

Figure 7.9. *Properties of a quantum system: the superposition of quantum states in a single qubit, then the entanglement between two qubits forming a single entity and teleportation principle, the reliable transmission of quantum information between two distant places symbolized by two people called Alice and Bob (adapted from [BEN 13])*

7.2.2. *Experimental devices*

To construct a quantum computer, it is necessary to start by producing a processor with logic gates like circuits of transistors in classical computing. The experiments were developed on three main types of physical supports with long coherence times. These are superconducting nanocircuits, trapped ions and qubits based on electron and nuclear spins in a material (note that other approaches exist as using polarized photons). They rely on the electron or core spin, its transport and detection, and associated technological advances such as optical manipulation. This is an extension of the field of spintronics in conductive materials that we introduced at the end of Chapter 4, even manipulating individual spins (i.e. nanospintronics).

7.2.2.1. *Superconducting loops*

They are based on the Josephson effect (Nobel Prize in Physics of 1973); a magnetic tunneling effect of a pair of electrons and a quasi-particle behaving like a boson in a superconductor (see definition in Table 5.1). It is a loop comprising a junction with a superconducting–insulator–superconductor multilayer structure, where the tunnel effect is controlled by a magnetic field or an external electric field. It gives rise to a continuous or alternating electric effect, which makes it possible to produce a detector, forming an ultrasensitive loop detecting magnetic quanta. It is called a superconducting quantum interferometric device that can be used as a detector or as a magnetic memory. Conventional materials are based on lead and niobium, metals and alloys transiting toward a superconducting state at very low temperatures, which involves the use of liquid helium as a refrigerant liquid. Initially, this high-performance technology was developed to create logic circuits and was tested in the laboratory without exceeding the experimental stage of a prototype computer at IBM [MAT 80]. This approach was resumed by creating a superconducting quantum bit formed by a double magnetic loop controlled by magnetic induction coils (Figure 7.10). At low temperatures, the electric current flows freely in one direction and produces a quantum of magnetic field, which is coded 0 or 1. This circuit is subjected to a quantum annealing effect operating by a tunnel effect, which makes it possible to change state and perform binary operations. A machine based on this principle is currently marketed by the company D-wave [BAL 14].

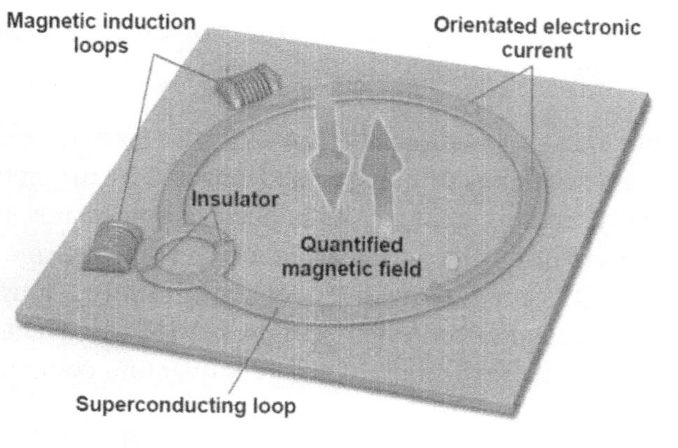

Figure 7.10. *Example of a superconducting quantum bit formed of a double loop of a superconducting material and controlled by two magnetic induction loops producing an upward or downward magnetic field (state 0 or 1) (adapted according to [BAL 14])*

7.2.2.2. *Trapped-ion technique*

From a practical point of view, the manufacture of a qubit is contradictory because they must be isolated from each other and their environment, as we have pointed out. Nevertheless, the interaction of the qubits with one another in the system is essential for the exchange of information. The ions trapped at very low temperatures are, for example, calcium atoms from which several electrons have been removed. Being unstable, they are manipulated in a vacuum using electromagnetic fields generated by lasers or using nanoscopic electrodes. This is an example of the technically delicate work of Wineland, which has allowed long coherence times to be obtained, permitting the coding of quantum bits and the realization of a logic gate [HAR 15].

7.2.2.3. *Control of spins in a material*

The spin of an electron or a nucleus can be at the origin of a quantum operation if its coherence time is long enough [PFA 14]. This is the case with a single diamond crystal doped with nitrogen: indeed, it is an extrinsic semiconductor with a level of impurities inside a wide forbidden band [DEL 11a]. This level of intermediate energy is accessible; when the electron falls back into its fundamental state, it emits a monochromatic photon by fluorescence. Upon doping with a color center, a nitrogen atom is in the substitution position in the lattice, associated with an electron spin, which is an environmental probe. Its dynamics are controlled by a gold circuit traversed by an electric current, producing photonic radiation that influences quantum states and their teleportation. The principle of such associated teleportation is shown in Figure 7.11.

Teleportation occurs photonically between two remote electron spins encoded on the nuclear spins of nitrogen and detected by magnetic resonance [PFA 14]. Finally, in this context, with nuclear spin states that are less sensitive to the environment and located on a magnetic molecule, their quantum character is preserved over sufficiently long durations (approximately 1 second at room temperature) [VIN 12]. Then, thanks to a molecular transistor, the realization of a purely quantum communication can be accomplished.

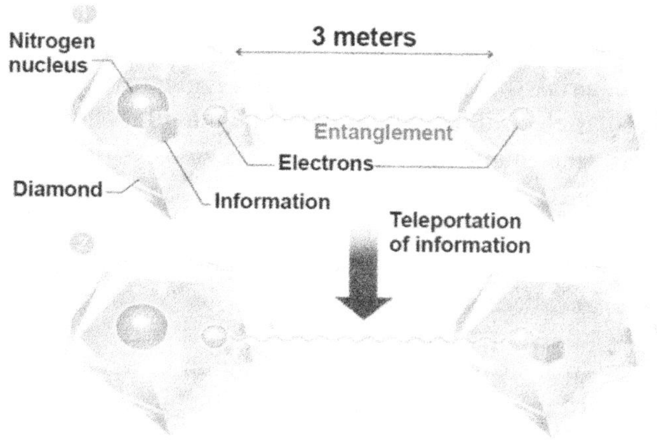

Figure 7.11. *Representation of unconditional teleportation in a doped diamond crystal. Information on the quantum state of a nitrogen nucleus is associated with the electron spin present; by entanglement, it is correlated to another electron spin in a second diamond located a few meters away, and by teleportation, the information (small gray cube) is transferred to the nuclear spin state (adapted from [PFA 14])*

7.2.3. *Information, thermodynamics and quantum chaos*

In Chapter 5 (see section 5.3.3) we mentioned the Landauer principle, which states that the irreversible erasure of classical information stored in a system requires work, dissipates energy and costs entropy. For ideally reversible quantum information, Landauer's formulation is called into question and may have a different meaning for the concept of entropy [DEL 11b]. These authors show that, under certain conditions linked to the phenomenon of entanglement, the erasure of data would cause a cooling effect instead of the emission of heat, which could result in a negative variation of the so-called conditional entropy of the system. This approach is linked to the possibility of a reversible transformation of the information associated with the presence of Maxwell's demon [BEN 03].

Finally, in a quantum system in dynamic regime beyond instability thresholds, one must quote the concept of quantum chaos, which establishes the connection between a classical deterministic system and the quantum nature of the particles present. This theory, based on the comparative analysis of their trajectories in classical and quantum mechanics, is currently being investigated [SOK 11].

7.3. State of the art: nanomaterials and quantum electronics

New nanomaterials have emerged and have given rise to the evolving fields of molecular electronics, spintronics and quantum photonics. Thus, we have developed the main functionalities previously defined from new formulations, created by chemists, often of an organic or organometallic nature. They offer possibilities in the quantum field, notably in molecular electronics and magnetism, but also in nanophotonics associated with artificial structures called metamaterials.

The tendency to focus research on the isolation of a single molecule and the manipulation of quantum particles must be included in a more complex, spatially self-organized and own logical set. This lack of architecture can pose problems and could be a technological barrier in contrast to that of silicon technology, which has been developed and improved for more than half a century.

As Feymann introduced, to store and read quantum information, a global device capable of manipulating a qubit according to certain rules is required. For this, the physical supports developed specifically to define a qubit presenting a long lifetime are presented. We have recalled and compared the main advantages and disadvantages of different technologies, noting that their stage of development is completely different. Ultimately, the computing capacity should far exceed that of conventional computers, but current realizations only concern a few tens or hundreds of qubits enabling a simplified processor to be realized. Their computing speeds are higher but their use requires the development of new algorithms. Finally, we highlight a particular point concerning the encryption of data, deemed inviolable for quantum systems [ROM 14].

From a technological point of view, the various routes involved require sophisticated approaches such as very low temperatures, a very high vacuum, high-performance nuclear imaging techniques or photonics to develop an optical computer. Finally, quantum computers are fragile, sensitive to the environment and therefore to observation. The reliability of logical operations is an essential factor to control. A flexible organization of small network-connected modules is currently emerging as a possible development pathway [MON 16].

8

Living World, Biomaterials and Biosystems

The passage from the inert world to the living world is a fundamental question that will be continually discussed in this chapter. We will consider living systems as a continuity of the physical universe, as described in the previous chapters. This description will rely on the global thermodynamic approach of closed or open systems, which was introduced in Chapter 2. In this context, we will review the energy transformations and information storage in living environments by looking at some simple examples in molecular and cellular biology. Indeed, our aim is not to explain the nature and evolution of the living world, but to present it as an ideal example for the development of biotechnologies. This has been in development for more than 20 years with the concept of biomimicry, with nature inspiring technological innovations at many levels [BYE 02].

Therefore, the concepts of energy and information in thermodynamics will be addressed in part one. In doing so we will distinguish, in this biological order, the functional aspect from the structural one at the cellular level. The emphasis will be on the functional organization, which we have already introduced for materials, which will become key in an initial step toward the concept of complexity, linked to the level of description of the systems. In fact, five levels are essentially present: molecular and cellular biology, then specific organs and individuals, with reference to the physiology of an organism, and finally population dynamics. These last organization levels, individuals alone or grouped in an ecological context, will be dealt with in Chapter 9. All these levels are characterized as systems with tangible limits or walls in accordance with the definition given in Chapter 2.

In part two, we will focus on the mimicry or inspiration of these biological transformations and conversions when creating biomaterials and biosystems, involving matter, energy or information. For this purpose, we will describe the main classes of artificial or hybrid systems, that is, mixed living-inert systems, and their current developments in relation to the thermodynamic classification already introduced in Chapter 3. Finally in section three of this chapter, we will discuss the increasingly sophisticated advances in man-made systems. The new developments resulting from this, in the context of thermodynamic models in an out-of-equilibrium situation, will be also be discussed.

8.1. Living systems and energy balances

The problem of understanding how a living organism can be described using the laws of physics and chemistry has been a recurring subject for more than a century, as we pointed out in Chapters 1 and 2, highlighting the fundamental role of morphogenesis. One of the authors who has approached the subject is Schrödinger, one of the fathers of quantum mechanics (Nobel Prize in physics 1933), who began to establish links with thermodynamics, in particular the entropy function. In his book published in 1944 [SCH 44], he linked the concept of evolution of entropy and that of information for the transmission of hereditary characteristics. We shall discuss the essential characteristics of living systems by confining ourselves to the cellular level in plants, and then animals. We then provide a thermodynamic approach to show that more specialized cells allow efficient energy conversions. Finally, we shall review sensory functions and energy yields, which make it possible to acquire and memorize information in animals, particularly in humans, which will be the subject of Chapter 9.

8.1.1. *On the definition of the living world*

A living entity can be characterized as an autonomous and evolving unit defined at two levels:

– an individual organism where its minimum structure is defined by the ability to remain alive, that is to preserve, self-regulate and reproduce. These fundamental functions are achieved through the exchange of matter and energy with the external environment and a set of biochemical reactions characterized by metabolism;

– as a population of living entities with a capacity to evolve through the process of natural selection by a replication of the genetic material with an almost exact transmission of information assuring the evolution of species (Darwin's theory).

To take a brief definition given by NASA: life is an autonomous chemical system capable of following a Darwinian evolution. The associated semantic problem is the transition from the inert world to the living world, which is not simply a self-organized system. A significant example is that of prebiotic systems, initiated by Oparine's work, followed by Miller's experiments on the primordial soup, inducing the synthesis of amino acids and polypeptides, precursors of proteins [BAR 13]. Indeed, the emergence of a reaction cycle in the models of Eigen (Nobel Prize in Chemistry 1967) and Schuster for the duplication of a molecular sequence is an approach based on spontaneous self-organization with the acquisition of information giving rise to the genetic code [EIG 77]. Current research focuses on the key role played at this level by RNA. However, our aim is not to analyze the origin and evolution of living species, from birth to extinction, but to simply consider the behavior of a living organism as an open thermodynamic system.

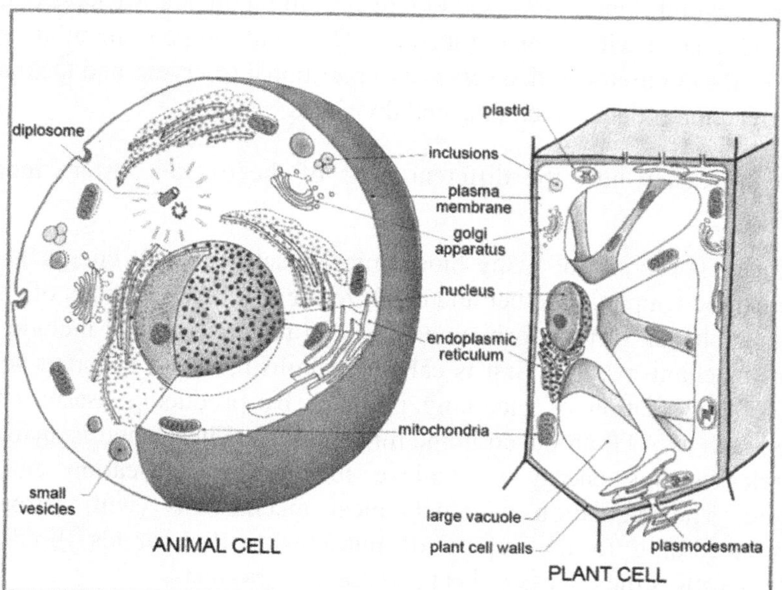

Figure 8.1. *A comparative diagram of an animal cell and a plant cell showing the main constituents (adapted from [CAI 06])*

This modeling will be initially applied to specific living eukaryotic cells, which are the most evolved multicellular entities (the three major classes of cellular structures being bacteria, archaea and eukaryotes). In order to consolidate this, Figure 8.1 shows the universal diagram of an animal cell and a plant cell [CAI 06]. These cells present different morphologies but similar organizations: a skeleton and a plasma membrane of a different nature, allowing selected exchanges with the exterior, various organelles such as the mitochondria and plastids or the Golgi apparatus, which each have specific functions, and finally the nucleus where the genetic information is located. A cell is therefore a heterogeneous medium with specialized compartments that can be assimilated to interactive subsystems. We will consider it initially as a global system behaving in a homogeneous way in interaction with its environment.

8.1.2. *Thermodynamic model*

Here, we begin with some definitions of an open thermodynamic system exchanging matter and energy with the seat of internal biochemical reactions. The situation is similar to that of a chemical reactor operating continuously, in and out of equilibrium regime (see Figure 2.5). In Figure 8.2, we have outlined such a cell reactor with its main functions. The inputs come from an identified source or the environment that serves as a nutritional reservoir, and the outputs are the reaction products including cell division.

Let us review how the different parts of the overall system function [CAI 06]:

– a cell is the site of many biochemical reactions as an energy source, which can be coupled together and catalyzed (enzymatic reactions of certain proteins). The set of transformations is called metabolism and includes two distinct mechanisms. The first is called catabolism, which accounts for the chemical degradation of incoming products by producing usable energy blocks, such as ATP produced in the mitochondria. The second is anabolism that releases this energy to achieve functioning, edification and cell structure. The regulation of biochemical mechanisms (with retroactive loops), according to the supply of nutrients (carbohydrates, lipids and proteins), causes the release of heat and reaction products;

– the cell wall, a plasma membrane and a cytoskeleton that ensures shape and movement form a defined boundary zone with the environment. The membrane, according to its composition and structure (lipid bilayers with ionic channels), regulates and selects mass exchanges of neutral or ionic species and the electrical signals received. It is specific for each type of plant or animal cell.

By taking the definition of the intensive parameters already explained in Figure 3.1, the possible transfers of matter and energy are in agreement with the associated intensive locally defined variables. They are controlled by the local chemical potential and an electrical membrane potential related to the concentration gradients of certain mineral ions (K^+, Na^+, Ca^{++}) and to the distribution of the intra- and intermembrane charges due to the presence of the channels, respectively. This phenomenon of osmotic pressure is related to the electrokinetic effects presented in Chapter 2 (see section 2.2.1).

– The cell division cycle is the division of a mother cell into two identical daughter cells, a mechanism that allows the transfer of genetic information via DNA. An essential characteristic of these systems is their specificity and their autonomy, presenting their own biological rhythms with an internal clock of genetic origin to control and transfer this information [KLI 14]. Note that the appearance of cancer cells is linked to an uncontrolled multiplication of cells. A change in the biochemical mechanisms of metabolism (Warburg model, Nobel Prize in Physiology 1931) may be one of the causes of cell proliferation linked to an increase in the rate of cell division. Reprogrammed anabolism is considered to be a deregulation for the production of the cancer cells at the origin of their proliferation and a modification of the cell cycle [DEB 16]. On the contrary, cellular senescence will result in the gradual disappearance of this division rate in a physiological process of aging.

From a thermodynamic or bioenergetic approach, all the transformations and uses of the energy are related to the metabolism rates and the associated yields. As we have introduced in Chapter 2, the concept of system evolution must distinguish two distinct regimes [PRI 72]. Close to equilibrium, the initial structure associated with entropy growth is maintained; this is the case for basal metabolism, which is always present, to preserve the vital functions. With increasing exchanges, beyond an instability threshold that must correspond to a bifurcation point, periodic structures inherent to nonlinear biochemical kinetics can be created: they result in an increase in the production of entropy observed in a dissipative system (see Chapter 9). Metabolism is linked to the increase in anabolism, thereby allowing the phenomena of energy consumption and reproduction by controlling the cell division. The transfer of genetic information takes place via DNA and RNA as well as the production of associated proteins. For this, the living organism, unlike an inert system, must be causally closed, that is to say, initially possess this instruction. Such an organization of matter might correspond to an overall decrease in entropy by considering the system along with its environment if the positive internal creation of entropy should be lower in absolute value than the term resulting from entropic fluxes (see equation [2.6]) [PRI 72].

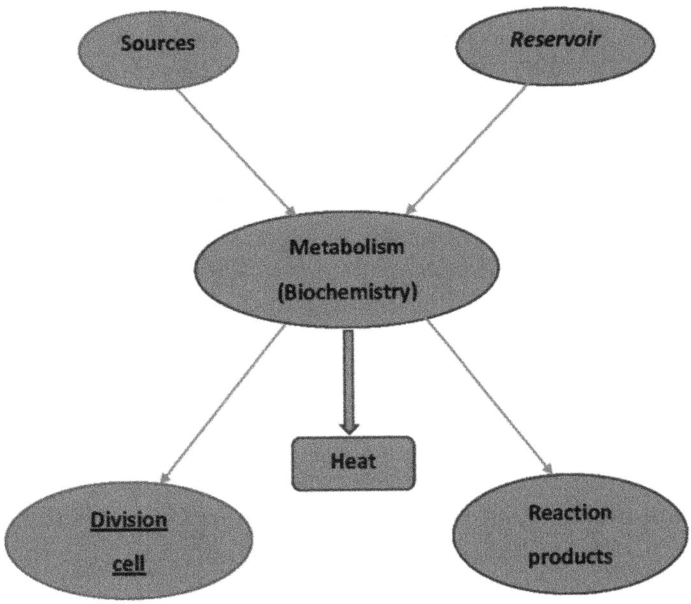

Figure 8.2. *Open thermodynamic model to describe the functioning of a living cell*

8.1.3. *Conversion and storage of energy*

We examine the production of main types of energy in the plant and the animal world. Photosynthesis in plant cells is the initial stage that transforms radiative energy from the sun into chemical energy. In animals, we will only discuss two key examples concerning specific multicellular cells. These are the conversion of this chemical energy into electrical energy by nerve cells or neurons, and the transformation into mechanical energy in the cells of muscle fibers. Finally, the information provided by a sensory organ will be introduced in this context.

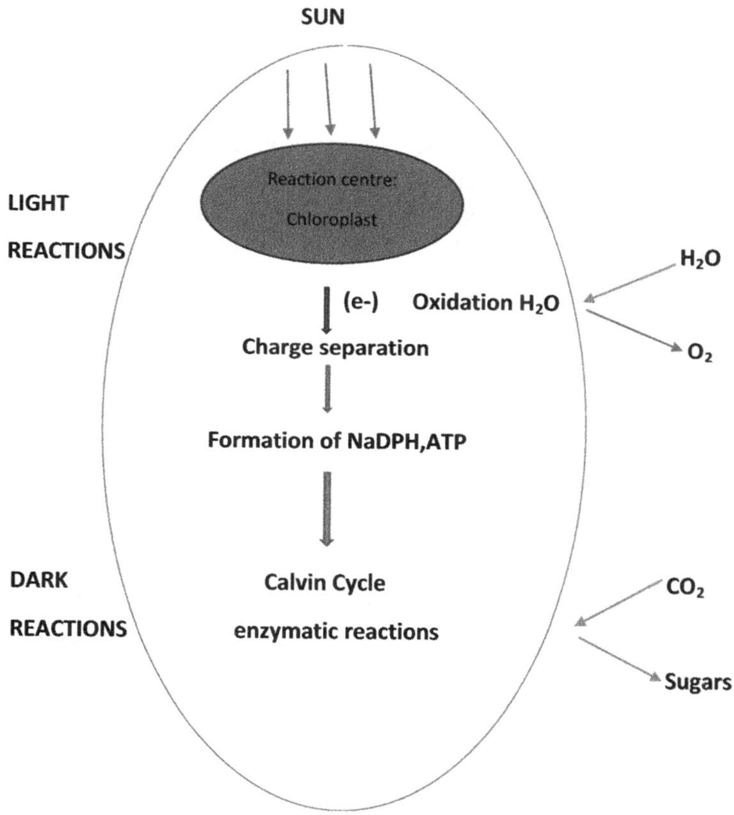

Figure 8.3. *Schematic diagram of photosynthesis in a plant cell*

8.1.3.1. *Natural photosynthesis*

We will describe the main steps of a complex mechanism that creates energy in the form of sugars from water and carbon dioxide. This is the fundamental mechanism for living systems (plants, algae and certain bacteria), which is the source of many technological developments. To consolidate this, in Figure 8.3 we show the two important stages of this mechanism. The photophysical part is the absorption of the visible light by a chlorophyll pigment subjected to a particular molecular set, known as the reaction center, where the separation of the charges occurs and an electron is released. This loss in the reaction center is compensated by the decomposition of water, which releases oxygen to the outside and produces protons. The resulting reactions lead to the formation of NADPH (protonated nicotinamide adenine dinucleotide phosphate) and ATP molecules, which can store and transport this chemical energy used in the oxidation–reduction reaction (transformation into NADP) and formation of ADP (adenosine diphosphate). These molecules are then used in a second step: the Calvin cycle. During these dark reactions, carbon dioxide is used for the synthesis of sugars (glucose in general) due to catalytic reactions (enzyme RUBISCO). The fixation of atmospheric carbon dioxide by plants or algae is a crucial part of the natural carbon cycle. Finally, the overall energy efficiency of conversion is estimated to be approximately 5%.

8.1.3.2. *The case of animal cells*

Plant cells, which are autotrophs, capture and transform solar energy, and this involvement of energetic molecules contributes to the functioning of animal cells. Plant cells are at the base of the energy pyramid in animals. Different animal cells result in functional specificities in multicellular organisms, with various energy conversions. Chemical energy is stored in the mitochondria of cells and used enzymatically. We will mention the most convincing cases from these biochemical transformations [BAR 13], such as the phenomenon of respiration with the oxidation of sugars or anaerobic fermentation:

– transformation into heat through global exothermic enthalpy of biochemical reactions that also contributes to the establishment of an internal or body temperature in humans and mammals;

– transformation into electrical energy with neuronal cells; these are highly specialized cells constituting a body, cytoplasmic extensions and dendrites with a terminal axon, which allow communication with other cells and form a neural network (Figure 8.4). The excitability of cells creates electrical modifications and the propagation of electric charges in these cells from a stimulus. Ionic conduction occurs during the transmission of a membrane action potential called a nerve impulse. It spreads along an axon insulated with a myelin sheath and has a synapse at its end, which is an active interface. The exchange of information between a neuron and a target cell occurs through the chemical intermediation of a specific neurotransmitter [CAI 06]. These neurotransmitters, e.g. hormones for the endocrine system, are the chemical messengers in a living environment;

– the transformation into mechanical energy can occur during the initial phase of cell division or during muscle contraction. In skeletal muscles, this phenomenon of motor movement passes first through a sensitive nerve fiber, a reflex center (the spinal cord) and a motor nerve fiber. Electrical conversion occurs, followed by signal transfer by a neurotransmitter and contraction of the muscle fiber, where we find a biochemical mechanism using ATP (see Figure 8.4). It corresponds to various types of innate or acquired reflexes in humans, including heart muscle. It is a complex contractile machine caused by a system of motor proteins in the cytoskeleton that transform chemical energy into work [WOL 01].

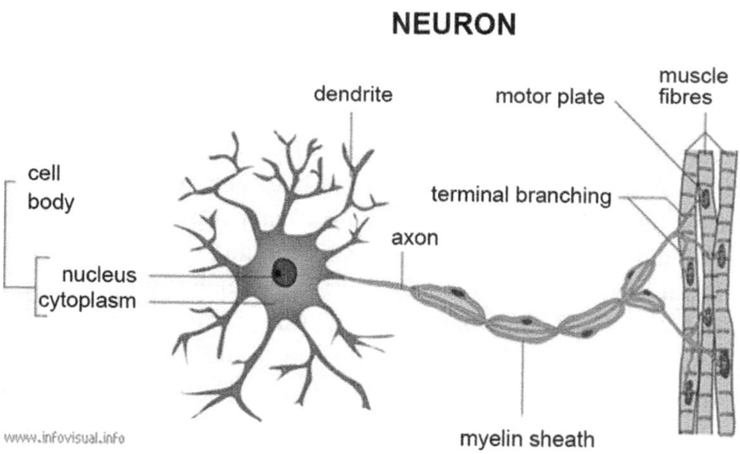

Figure 8.4. *Diagram of a nerve cell or neuron and its constituents, allowing the transport of an electrical signal, specifically, in the case where an axon can excite muscle fibers*

8.1.3.3. *Sensory receptors*

Sensory organs, typically the five senses, are made up of specialized cells that detect external stimuli. They provide information that must be analyzed and stored in a memory. As we have just seen, it is the role of the nervous system and the spinal cord that ensure the transmission of electrical signals leaving or going to various parts of the brain. Thus eyes, ears, noses, tongues and fingers are natural transducers. They are sensitive to light (sight), pressure and heat (touch), chemicals (smell taste) and sounds (hearing). They transform a physical stimulus into an electrical response that, via the neurons, goes to the brain to be analyzed and ultimately stored.

Vision, i.e. shapes and colors, is a good example with the eye similar to a photographic optics camera, whose detection is assured by visual cells in the retina. An electrical potential is then created and detected and this encoded information is transferred to the brain. This information for visual recognition of the forms is sent to a part of the brain. It is analyzed and stored as a particular type of memory as we find in neuroscience [CAI 06]. These sensory structures allow organisms to respond to their environment, which is no longer simply a source of energy and matter, but also information.

8.1.4. *Operation of a cell reactor*

A living cell is the basic unit in all living systems and by extension their assemblies and the associated natural cycles too. As discussed in section 8.1.2, a cell is identified with an open homogeneous thermodynamic system where the level of matter and energy exchanges and their implications (see Figure 8.2) can be analyzed. From a biochemical perspective, a description in classical thermodynamics based on the Gibbs function and the enthalpies of reaction makes it possible to account for energy exchanges and storage in a living cell [PRI 72]. Thus, enzymatic activity allows a cell to respond to changes in its environment, that is to say, to the impact of the level of energy resources that it tends to regulate by feedback loops. These are genes that respond differently and define changes in metabolic level and the concept of a threshold behavior change as shown by a numerical simulation to calculate the flows exchanged in a cell [COV 14]. Nevertheless, for an active metabolism, the thermodynamics of living systems, far from equilibrium beyond a bifurcation, is at the origin of oscillatory dynamics and phenomena of spatial segregation [RIC 99]. The temporal organization of metabolic cycles [KLI 14] with the appearance of periodic oscillations that can be disrupted [DEB 16], possibly followed by a chaotic situation, will be discussed in Chapter 9.

8.2. Biomaterials and biosystems

All research, using living organisms as a model, can be classified according to three approaches:

– making a copy of living systems by mimicry, taking inspiration from the forms and functionalities observed, to obtain similar or improved behaviors;

– introducing a natural part in mixed devices, such as sensors or transducers, in order to improve its overall efficiency, for example, by using an enzymatic catalytic reaction;

– modeling by calculation or biochemically constructing a gene or even a cell within the context of synthetic biology.

In this context, biocompatibility is defined as the tolerance of cells to various materials. The behavior of a living environment in the presence of a material is fundamental. The characteristics of the interface created were introduced in Chapter 3 (see section 3.1.1). At this interface, a possible chemical reaction can occur: the material is said to be biodegradable. This surface effect depends very much on the nature of the implants and the size of any particles present (challenges in nanotoxicology).

We will address these three aspects by discussing biomaterials and biosystems, classified following functional purposes. In order to do this, we use the classification used in Chapters 3 and 4, involving the different forms of energy, their conversion and the processing of information.

8.2.1. Morphogenesis and biomimicry

The evolution toward complex forms in an open system was introduced in Chapter 2 and developed in the context of thermodynamics far from equilibrium: it is morphogenesis that links inert matter and living matter. We will focus on the use and copying of natural forms, in particular, their architecture at different scales to obtain hierarchized materials. The aim is to exploit or even to improve a particular natural property: to illustrate this, we will provide some classical examples.

8.2.1.1. *Formation of self-organized patterns*

The observation of living systems is what inspires the creation of advanced materials by biomimicry [BYE 02]. These include, in particular, hybrid materials that are composed of an organic and an inorganic part; they are manufactured by techniques known as soft chemistry (the sol-gel process for instance, see Chapter 6). An example provided by Livage [LIV 11] is that of monocellular or diatomaceous algae formed from silica shells that have a specific porosity. These are living photonic crystals that optimize light harvesting and photosynthesis. Another good example is that of mother-of-pearl shells, a very tenacious material that breaks with difficulty, due to complex hierarchical structures. It is made up of aragonite (calcium carbonate) in the form of platelets welded together by proteins in oriented columns. This natural lamellar material has been reproduced in the laboratory to obtain a ceramic that combines a high Young's modulus with good mechanical toughness [CHE 14].

Many developments are based on the determining role of surfaces and interfaces with particular geometric and functional characteristics. There are two examples relating to bonding and wetting that are ideal interfacial properties. Blue mussels secrete an adhesive that allows them to cling to rocks; it is a peptide that ensures wetting, and then through oxidative cross-linking ensures cohesion. Biocompatible surgical adhesives have been produced based on this principle. Another example is the gecko, an insectivorous reptile that is a very good climber. The soles of its feet are textured and present a hierarchical structure terminating in keratin hairs, which provide an exceptional reversible adhesion. This topographic characteristic or roughness, based on Van der Waals' physical interactions, was used to control the wetting transition on a given surface [BON 01].

8.2.1.2. *Fibers and hierarchical structures in composites*

Plant and animal species are made up of natural macromolecules that form fibrils, which are assembled by successive levels forming hierarchical organizations. Let us mention some prototypes; in the plant world with cellulose, a polysaccharide present in the cell wall of microfibrils. From multidimensional successive layers, a fibrous system is created constituting the microstructure of the wood. An illustration of the internal organization of cellulosic chains at the nanometric level is shown in Figure 8.5. This organization leads, at larger scales, to the formation of plant cell walls and the mechanical qualities of wood due to an oriented distribution of fibrils. A similar oriented organization is found in certain components of animals such as hair, based on keratin and tendons, which are made of collagen [VIN 02].

There are also natural fibers of inorganic origin, such as asbestos, or plant origin as cotton or linen, and even animal origin such as spider's thread. Silk thread of 10 μm in diameter is renowned for its optimum mechanical strength. Natural filamentary or oriented materials, such as wood or bone, have interesting mechanical properties that have been copied by engineers to make synthetic fibers and composite materials. These fibers are insulated, spun and can be used to make rods and fabrics, used as mechanical reinforcement. An analysis concerning the elastic properties has already been discussed (see Figure 3.2) and we will only describe their multidimensional character.

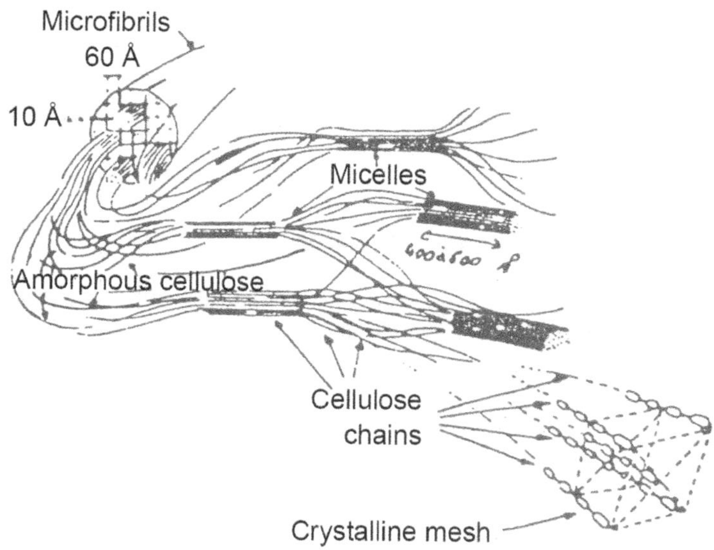

Figure 8.5. *Drawing representing the interior of a cellulose microfibril at various scales, embedded in lignite, another component of wood*

Synthetic fibers were the prerogative of polymer chemistry since the discovery of nylon. Some polymers of high mechanical strength include Kevlar, polyethylenes or various types of carbon fibers obtained from organic precursors, which include a variety of stiffeners adaptable according to the requested variables of size and duration. In a composite material, consisting of a fibrous reinforcement inside an organic or inorganic matrix, different spatial scales must be examined. It is truly an artificial hierarchical structure that is optimized by means of predictive calculations in mechanical modeling [DEL 06]. The analogy with natural systems can be extended to the nanoscale level with nanotubes and spun-oriented carbon nanofilaments (see Figure 6.8) that copy the organization of cellulose, as shown in Figure 8.5. They are used to produce nanocomposites, which will exhibit specific mechanical characteristics due to an increased interface, given for a unit volume. A recent example is that of orientated graphene nanoplates in a polymer with a particular combination of mechanical, electrical and optical properties [LIU 16]. These dispersions are also used to restrict the propagation of fractures under nonlinear stress conditions, such as in concrete or in the case of tires reinforced with carbon blacks or silica particles. Finally, in this context concerning structural materials, their use in the human body must be mentioned, for example, as prostheses and implants as long as their biocompatibility is good.

8.2.2. *Biodetectors and similar functions*

In this section, we shall group biosensors and other biological detectors in the field of analytical chemistry. The behavior of different devices is examined at different scales down to the genetic traits with the use of DNA microarrays. Finally, encapsulation methods and drug vectors complete this presentation.

8.2.2.1. *Biosensors and interfaces*

A biosensor is an analytical tool combining a biochemical component (enzyme, DNA, etc.) that undergoes molecular recognition and transforms it into a measurable optical or electronic signal. Among all the fields of application, an emerging sector is the development of *in vivo* implantable biosensors [MAR 08]. The analysis of markers and biological constants in a patient is a significant advance, such as a glucose biosensor for monitoring diabetes. In Figure 8.6, an enzyme, a protein acting as a catalyst by means of a cofactor (active site known as flavin adenine dinucleotide) and a carbon nanotube are covalently bonded and connected to an electrode, which detects the electrical signal. This type of biosensor with a "living-inert" interface is increasingly generalized, for example in neurosciences, where a neuronal interface with a malleable implant can be created [MIN 15]. This interface can be optical to avoid a biodegradable surface effect. Due to a molecular switch controlled by a light irradiation acting as a stimulus, communication between neurons could be directly established [LEM 13].

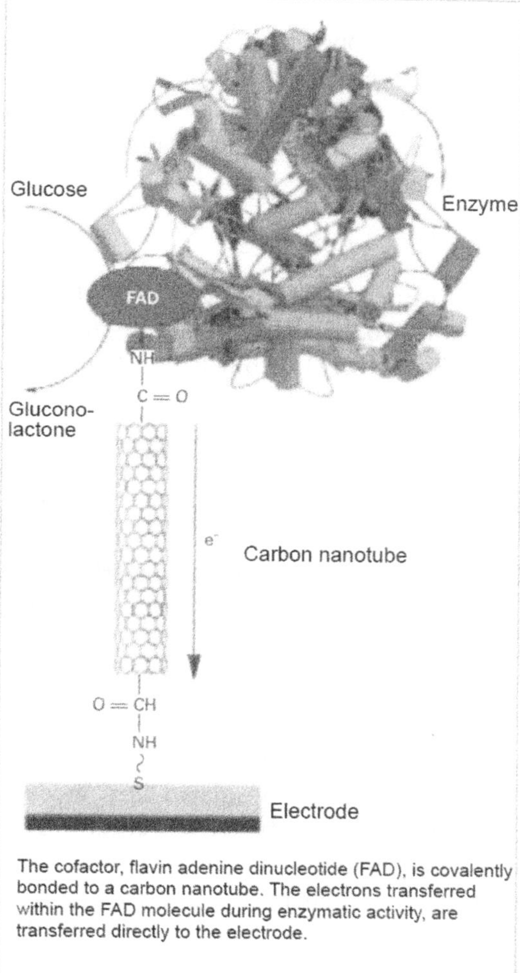

Figure 8.6. *Operation principle of a glucose biosensor using a carbon nanotube as a molecular wire (adapted from [MAR 08])*

8.2.2.2. *Microelectrodes or artificial synapses*

Electrochemical methods used as analytical tools are very efficient. Converting chemical information into an electrical or optical signal or a combination of both (e.g. electro-chemiluminescence) is a resource-rich pathway [MIO 15]. Thus, an electrode placed in the vicinity of a living cell can give information on the local changes in chemical species. This artificial synapse method allows us to study the mechanisms of oxidative stress in a living cell [AMA 11]. This approach is an example of a necessary interface between the biological environment and molecular electronics for implantable systems. This interface is characterized by the transition of an ionic natural conduction toward that of charge carriers of the solid-state physics, electrons or holes.

8.2.2.3. *Biochips and DNA networks*

These are devices capable of carrying out genetic analyses. The DNA molecule is formed of two helically wound strands with the four bases coupled in pairs (see Figure 8.12). A biochemical analysis makes it possible to detect the presence of a single strand of DNA by pairing it with a complementary strand acting as a probe, already fixed onto the chip. This is a miniaturization of biological spots aligned on a silicon support allowing a serial analysis due to a mastery of microfluidics. After optical detection and reconstruction of the data, the order of the nucleotide sequence can be determined or/and gene sequencing can be performed [CHÂ 05]. This operation is the first step in genomics, a field that we will discuss later in this chapter.

8.2.2.4. *Encapsulation and nanovectors*

In the field of nanomedicine, the development of therapeutic vectors is rapid. The active ingredient is encapsulated in a biocompatible envelope, which is introduced into the human body. The encapsulation is carried out in hollow structures such as micrometric microcapsules made of porous silica. This is an extension of the soft chemistry technique described in the previous section; a simple example regarding diabetes is given in Figure 8.7 [LIV 11]. Pancreatic cells, derived from a healthy organism, are immobilized with alginate in silica capsules allowing the exchange of glucose and insulin. The silica shell prevents immune rejection, because it does not allow antibodies to reach the target.

This encapsulation system based on core-shell differentiation in nanoparticles has been widely developed, in particular for the vectorization of an active principle and the diffusion-controlled release through the wall. An alternative is to produce vesicle-forming liposomes from amphiphilic lipids or polysaccharides, which are stable and biodegradable nanometric colloids. They thus control the distribution kinetics of the drug. Furthermore, these nanovectors should be able to target diseased cells and tissues in the body. They were developed in particular to release *in situ* antitumor (cancer) or antiviral (AIDS) drugs [HOR 11]. This encapsulation and injection of the active ingredients into the organism must be detectable by medical imaging techniques. Recent advances have been made in particular on the identification of specific chemical signals to deliver the active ingredient to the target area [MO 16].

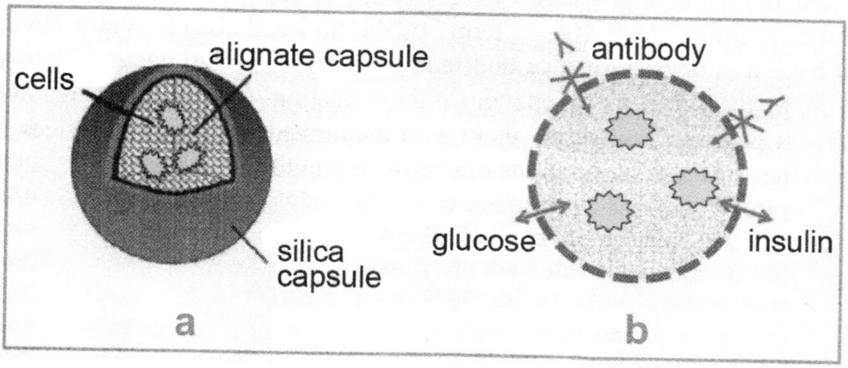

Figure 8.7. *Example of a hybrid encapsulation nanocapsule: a) the cells of the pancreas are trapped in a silica-coated alginate gel; b) the porous silica allows the molecules to pass but blocks the entry of the antibodies (according to [LIV 11])*

8.2.3. *Bioconverters and natural energy sources*

We are interested in devices that produce and store electrical energy from biological elements, i.e. biofuel cells. This approach can be generalized by exploiting the different natural forms of energy. At the beginning of this chapter, we went through photosynthesis (Figure 8.3). Our starting point is either a modified natural photosynthesis or an artificial approach. In addition, the exploitation of reactions involving water or carbon dioxide molecules by green chemistry is a complementary stage. These techniques produce secondary sources or vectors of chemical energy, namely hydrogen or organic fuels that can be stored.

8.2.3.1. *Biofuel cells*

The principle is directly derived from the operation of a fuel cell (see Figure 4.5). In the case of biochips, catalysis, which is carried out by noble metals, is replaced by redox enzymes or microorganisms. Thus, two types of bioelectrochemical techniques exist with biological catalysts of enzymatic or microbial origin [TIN 13] or even their associations [FIS 09]. In enzymatic biofuel cells, the catalysis of redox reactions is ensured by a particular class of enzymes called oxidoreductases. They are connected to an electrode and must therefore be immobilized; the intervention of a mediator ensures the transfer of electrons and an electrical current between the electrodes. A typical example is that of a glucose–oxygen couple (Figure 8.8), where glucose oxidation is provided by the electron-releasing glucose enzyme that reduces bilirubin [FLE 10]. In terms of the power delivered, these devices still have a poor performance and their lifespan is limited because they are linked to the problem of enzyme immobilization. The progress recorded makes it possible to imagine them as miniaturized implantable devices for supplying a pacemaker or an insulin pump *in situ* for example.

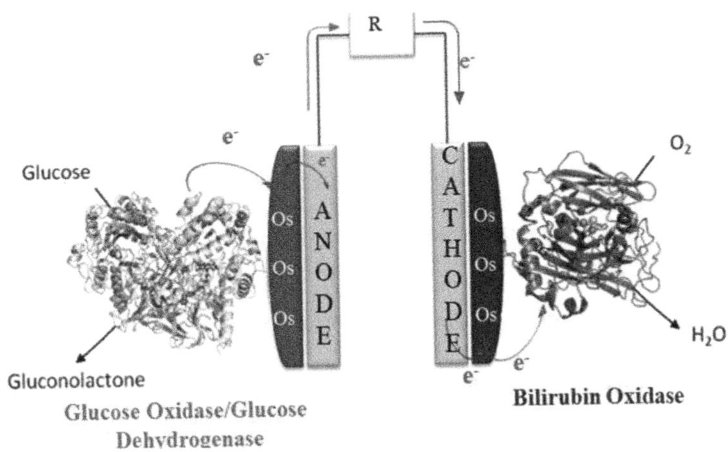

Figure 8.8. *Diagram of an enzymatic biofuel cell: the two electrodes are modified with biocatalysts operating autonomously in the same chemical solution(according to [FLE 10])*

Microbial batteries, which have been recognized for a century, are due to bacteria that produce electricity through the natural catabolism of sugars [TIN 13]. The degradation of certain organic molecules releases electrons collected by an electrode. The generation of an electric current by microorganisms, forming a biofilm attached to the electrodes, constitutes a fuel cell. Applications are limited by the control of microbial cultures, however its direct use in nature has revived this method. Indeed, this electrical current can be directly recovered in sediments of organic matter and even from vegetation. Some of the organic matter from photosynthesis is returned to the soil, where microorganisms are electrochemically active. They break down this material by generating electrons that can be collected *in situ* with a graphite electrode and thus continuously generate a small electric current. Research is underway to optimize this bioenergy production (see, for example, European program "plantpower") [HEL 13].

8.2.3.2. *Bioenergetic transformations*

The direct conversion to electricity is only a partial approach to the situation. Using living systems and drawing inspiration from photosynthesis, several pathways for the synthetic production of secondary energy have been developed. They make it possible to obtain, as a usable energy vector, electricity, of course, but also hydrogen and biofuels. These bioenergetic conversions are based on the fundamental mechanism of photosynthesis without taking into account photothermal effects (heat production). They are shown schematically in Figure 8.9.

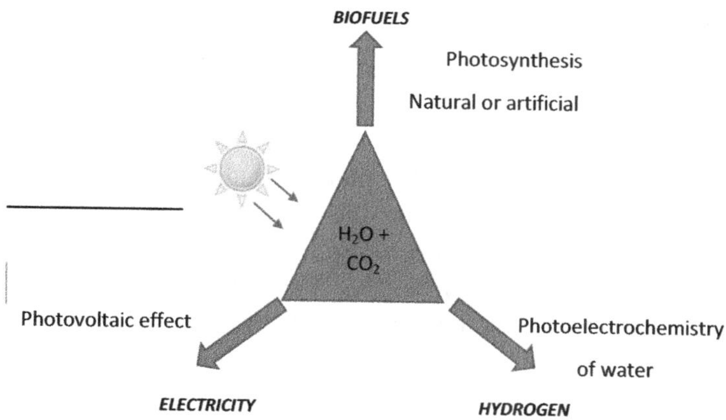

Figure 8.9. *The bioenergetic transformations associated with the different photoelectric conversion mechanisms available from solar energy (adapted from [RUT 05b])*

These approaches have two purposes, to initially improve the natural process of light absorption or to mimic it with chemical catalysts. Then in shade reactions, one must either involve the Calvin cycle with the use of CO_2 or carry out the decomposition of the water with the production of hydrogen.

For this, we examine the various secondary energy productions and then the specific case of carbon dioxide:

– electricity is generated by the photovoltaic effect, already described in Figure 4.1: broad forbidden band width semiconductors (silicon or perovskite) lead to higher yields than those obtained from chlorophyll pigments or carotene molecules. Note that with the biocells, the natural separation of the electrical charges is done using enzymes or microorganisms;

– biofuels are prepared from biomass: For second- and third-generation products, they come from the cultivation of inedible ligno-cellulosic plants and micro-algae, which offer a higher fuel yield [WIF 10]. It is necessary to distinguish the sector of vegetable oils for biodiesel and that of bioethanol obtained by the fermentation of sugars. We must add biogas to these sectors, largely based on methane and originating from the increase in recycling of agricultural waste. There is also the possibility of using solar energy thermally: thermochemistry makes it possible to decompose thermodynamically stable molecules of water and carbon dioxide at very high temperature and to obtain molecular hydrogen and carbon monoxide. Synthetic gas or syngas is a source of mechanical energy in an engine [SCH 14];

– the photolysis of water by a photoelectrochemical cell under the action of sunlight has been the subject of numerous studies. We will discuss some examples of the production of hydrogen made from genetically optimized microorganisms or artificial biomimetic catalysts. Indeed, the photodecomposition of water is possible due to metallo-enzymes known as hydrogenases, as those found in green algae [BER 13]. To improve the conversion with modified enzymes, the activation mechanism has to be optimized [GUS 09]. However, artificial photosynthesis is currently the most explored route. Its principle is based on the coupling between a photovoltaic cell and an electrolyzer inside the same system, inducing the production of molecular hydrogen and oxygen. It can work using platinum electrodes but cannot be used globally due to its rarity and price: it is necessary to find alternative catalysts. An operating diagram is shown in Figure 8.10, in which the three main functions are represented [GUS 09]. An antenna made up of an organic complex (porphyrin and fullerene) collects and transforms the radiative energy into electrochemical energy via a redox reaction. The oxidation of water occurs with inorganic catalysts (metal oxides), then with a second type of catalyst, the proton is reduced to dihydrogen. The challenge is obtaining a reliable device with a non-negligible yield by separating the production zones of the two gases.

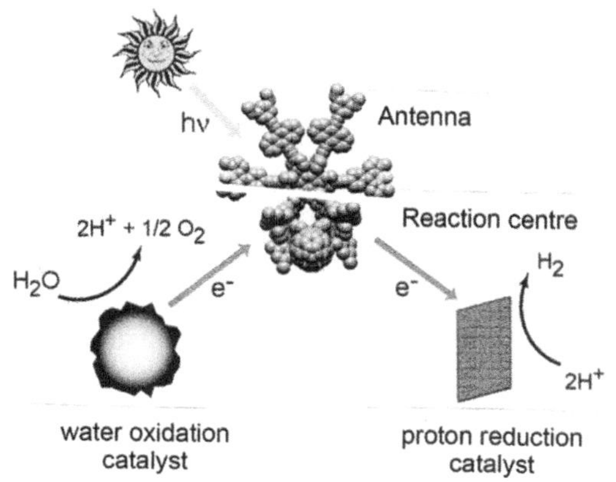

Figure 8.10. *Operation of an artificial light-harvesting device with the production of hydrogen (adapted from [GUS 09])*

Following this method, the most successful, but costly, project is known as "artificial sheet", where thin layers of silicon and various abundant metals are assembled to obtain a yield of dihydrogen, from 2.5 to 5% of solar energy received, according to the configuration of the device [REE 11].

– Assimilation of CO_2: A positive alternative to the capture and storage of carbon dioxide, which is the main gas involved in the greenhouse effect, is to recycle it chemically. CO_2 intervenes in the biogeochemical cycle of carbon, where it is fixed inorganically (CO_3Ca) or organically. In the latter possibility, different approaches are developed for energy recovery.

Among the attempts to assimilate CO_2 by organic synthesis, it is necessary to review the solar thermochemistry of the water and carbon dioxide mixture [WIF 10]. Other approaches have been proposed using electricity to initiate this type of reduction reaction. This is the case for methanization, discovered by Sabatier (Nobel Prize in Chemistry 1912) more than a century ago, resulting from the exothermic reduction of carbon dioxide by dihydrogen:

$$CO_2 + 4H_2 \longrightarrow CH_4 + 2H_2O \quad [8.1]$$

The molecular hydrogen obtained by the electrolysis of water reduces carbon dioxide and converts it into methane, in the presence of a nickel catalyst, with good efficiency [BRO 07]. This method makes it possible to store an excess of electrical energy transformed into chemical energy. It is, however, linked to the electrolysis of water, an energetically costly process, which is on hold until an effective alternative technology is developed.

Finally, the electrolytic reduction of CO_2 and its conversion to formic acid, which is then transformed into methanol, is an interesting prospect [ELG 13]. However, this recovery testing, which may have a positive overall energy balance in the context of sustainable chemistry, is not yet developed at the industrial stage.

8.2.4. *Engines, receptors and bionic robots*

8.2.4.1. *Biological engines*

In Chapter 7, we introduced molecular engines as a nanotechnological process. The transition to more efficient biological engines and their replication is a new step. They are composed of a set of motor proteins operating in a living cell to establish a chemical–mechanical cycle: an example is given in Figure 8.11 [BUS 05]. A tubular lipid membrane structure, formed in vivo, is pulled by motor proteins (kinesins), which can induce movement. The obstacle is to then introduce these biological engines into a device to create a molecular robot. This will include detection via a sensor, data processing and an actuator in accordance with the instructions provided [HAG 14].

8.2.4.2. *Sensitive receptors and robots*

To do this, we must develop interfaces to transmit sensory information by replicating the five senses as we introduced in section 8.1.3. Such is the case for making hearing aids or those for restoring sight. The sensory inputs must generate a set of stimuli on a sensor, transformed and analyzed by an artificial intelligence unit.

Thus, in the context of opto-genetics, an aquatic robot was created that simulates the displacement of a ray through wave-induced movements produced by light [PAR 16]. Even the delicate touch, i.e. sweeping a hand over a surface, inducing dynamic pressure, which is then transformed into a nervous signal, has been mimicked. A prototype of artificial skin, containing its main morphological characteristics, has been produced [PAR 15]. It is made of a flexible multilayer composite that is sensitive to temperature and pressure. It is essentially composed of a piezoelectric material that detects the static pressure and another ferroelectric material (spontaneous electrical polarity depending on T and P), which makes it possible to detect any relative variations. These new transducers with sensitive capabilities are examples of recent significant contributions to the creation of sensory robots.

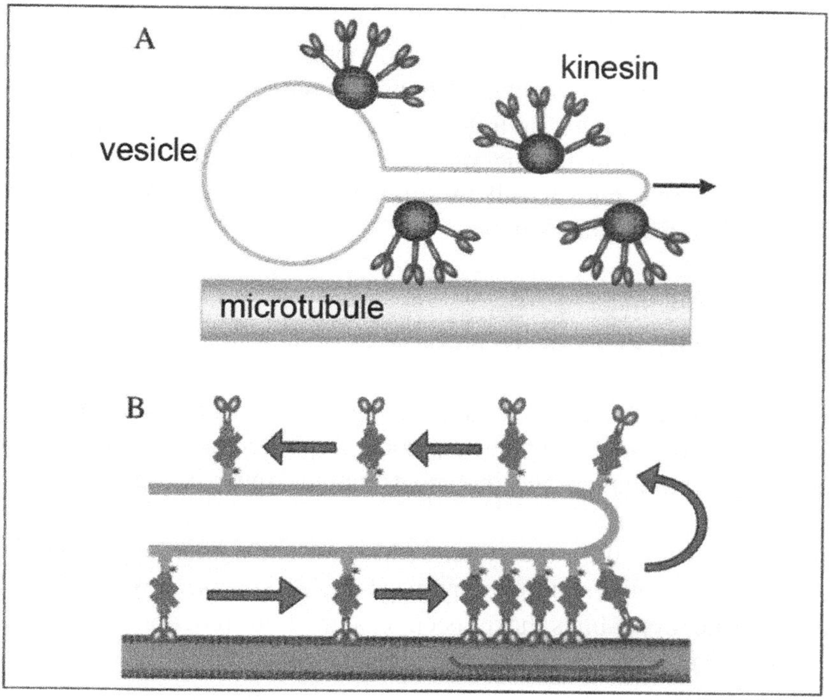

Figure 8.11. *Diagram of molecular systems used to draw a filament in the form of a microtubule: A) molecular motors are attached to the membranes of giant vesicles and move in the presence of ATP; B) molecular motors are directly attached to the lipid walls, and those at the tip exhibit traction by moving slower, creating motor accumulation at their tip (according to [BUS 05])*

8.2.5. *Bioinformatics*

In this section, we will describe three points mentioned at the beginning of this chapter. Firstly, the use of the computing capacity to simulate a living organism. Next, genetic information must be included in the information theory, already introduced in Chapter 5, by examining both genomics and biosynthesis. Both are initially based on the controlled coding of DNA (and RNA), which becomes the essential biomaterial as a digital memory medium. Finally, the concept of memo-computers, which is inspired by the neural functioning of the brain, is introduced.

8.2.5.1. *Numerical simulation of a living cell*

The dynamic modeling of a living bacteria cell is a significant advance [COV 14]. The introduction of all the biochemical processes and essential biological mechanisms allowed the modeling of evolution, growth and cellular reproduction, by functional transcription of the genes. The type of matter exchanges with the environment makes it possible to adjust the type of metabolism and functioning of this artificial cell. The objective is to model a eukaryotic human cell to simulate its behavior as a function of present stresses, fluxes and stimuli.

8.2.5.2. *Genomics*

The sequencing of genes and the analysis of their function and expression is the basis of an extension toward the living world. Designing a biological computer requires digital storage using DNA, and transmission by replication between cells, which is a good way to conserve it; a reading or sequencing technique via the biochip technique must be operational. The encoding of DNA (or RNA) as a digital memory was carried out by moving from a binary code (0,1) to a system coded formally in an upper numerical base, which corresponds to the four paired nucleotides (see Figure 8.12). For instance, gene assemblies have been designed to function as electronic components that can be inserted into a genetic circuit [DAU 13].

Biological systems are capable of copying, recording and manipulating information just like computers; it is possible to control biological communication with a cellular transceiver system. This is an alternative approach to current computer systems [DEL 17]. The advantage of this genetic information is the long conservation time that allows perennial preservation. However, as in any living organism, the repair mechanisms of DNA damaged by external damage or duplication errors must be analyzed and controlled (see Lindahl, Modrich and Sancar Nobel Prize for Chemistry, 2015). Although this allows the study of transmission mechanisms, it is not yet completely resolved for reliable storage and subsequent digital processing. However, this approach take into account the complications of genetic drift linked to mutations and the role of neighboring proteins in epigenetics when molecular mechanisms modulate gene activity.

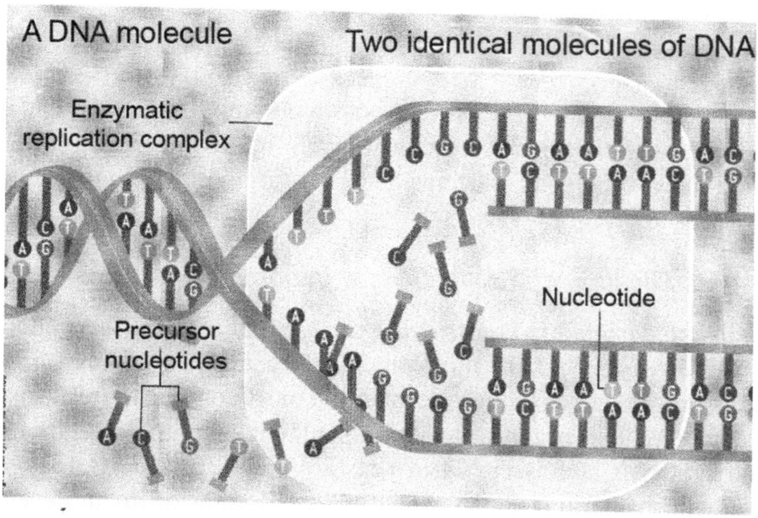

Figure 8.12. *The information encoded in the genome is the replication of DNA assisted by an enzymatic complex giving two identical molecules. The two strands are separated; then on each strand the complementary nucleotides are positioned in the sequence as provided by the ambient medium (A-T pairs for adenine and thymine and C-G for cytosine and guanine (according to [DAU 13])*

8.2.5.3. *Memo-computers*

Let us discuss the analogy with traditional computing, based on the read and store functions, to transmit and perform logical operations. Note that the architecture of a computer is based on the storage and sequential processing of information, which does not make it capable of data processing in parallel. It is totally different from that of the human brain and its assembly of neurons, which are the elementary operators for its various memories. A new concept is that of memo-computers, inspired by cerebral neurons, where logical operations and non-volatile data storage are performed together. For this purpose, a memristor is used, which is a passive electrical element that stores information, already described by Chua for the dynamics of coupled oscillators (see Chapter 3, section 3.2.2): it has been created experimentally and used [STR 08]. A biological transistor or transcriptor was also performed by modifying the genetic material in a DNA sequence [BON 13].

The development of these new components should make it possible to build memo-computers or neural computers [DEL 17].

8.2.6. *Biosynthesis*

Synthetic biology, which makes it possible to modify the genome or even to create an artificial one, is of considerable importance [KÉP 14]. When designing new biological systems and mastering their engineering, new ethical issues arise.

8.2.6.1. *Modification and creation of a genome*

This is the design and construction of essential biological components, a rapidly developing field. In this context, the synthesis of DNA has made enormous progress over the last decade; the most convincing result is that of the first unicellular organism with a synthetic genome. By digitization, machine synthesis and assembly in yeasts, Venter *et al.* [GIB 10] created a semisynthetic bacterium capable of dividing. Extension to more elaborate genomes of multicellular organisms is underway and the creation of an artificial cell, an autonomous and evolutionary living system containing transmissible information, is still a fundamental topic. The design and construction of a small genome for the functioning of a cell, with desired biological functions, is a declared objective [KÉP 14].

Since this first experiment, the modification of DNA can be carried out in several ways. A new technique is based on the presence of Clustered Regularly Interspaced Short Palyndromic Repeats (CRISPR), a sequence repeated in DNA. The association with a "CRISPR-Cas9" protein cuts the two strands at a selected position. It makes it possible to suppress or insert new sequences of the genome into several types of cells [JIN 11]. Treating genetic diseases becomes accessible, but alongside this is the emergence of issues caused by modifying gene expression and the heritage of a living species. Therefore, over the past few years, have been therapeutical objectives are balanced by industrial purposes and eugenic considerations, which induce ethical questions [TRE 15].

Finally, another approach is to increase the genetic alphabet by the addition of two letters [MAL 15]. The addition of two new base pairs of the genetic code of the DNA double helix (see Figure 8.12) leads to additional information. This synthetic DNA can replicate in a bacterium, which is very energy efficient and it could meet the growing needs of data storage. However, it also raises fundamental ethical issues.

8.2.6.2. *Biotechnologies*

Biotechnology has long been involved in diet due to the use of different yeasts. By introducing new biological functionalities, the real challenge is to extend the application fields in controlled bioreactors. Since the chemical synthesis of insulin, several bacterial syntheses have been carried out, like that of artemisinin, which is an antimalarial drug. This global trend is part of the expansion of green chemistry toward the production of biogenic nanomaterials from different microorganisms. Finally, bioinformatics, by realizing functional genetic circuits, could still become reality [KÉP 14].

8.3. Conclusion

In this chapter, we have discussed mimicry of living environments through the analysis of forms and functions [BYE 02] using the concepts already introduced in Chapters 1 and 2. For this, we started with autonomous morphogenesis and their operational mechanism, described energetically as being metabolism. In this section, we summarized the fundamental behavior of a living cell as an open biochemical reactor, operating at different levels, before extending it to more structured organisms. This leads us to the concepts of self-organization, hierarchy and rhythm, which introduce the dynamic concept of chemical and biological complexity [RIC 99], which we shall explain in the general conclusion..

With regard to the preparation of artificial materials and systems, we have shown how inspiration and the use of biological processes can open up new areas of applications. Indeed, living environments offer different solutions in relation to energy conversion and storage, such as in a muscle or a molecular motor from biochemical energy; we have only presented the major trends and some significant examples of biosensors and bioconverters. They generally operate close to equilibrium state, offering higher conversion efficiencies

This evolution of technologies has three distinct stages, ranging from bioinspired mimicry to mixed inert-living systems and finally integrated biology. The latter, based on bionics and synthetic biology, opens unexplored paths to create new fields of applications. Let us conclude by noting that the boundary between the inert and the living world becomes more and more blurred the concept of artificial life, thus introducing ethical issues.

9

Extensions to Living Organisms and Ecology

Non-equilibrium thermodynamics have allowed us to describe the varied responses of materials in physicochemical systems and extend them to living environments. New biological functions emerging from dynamic interactions between molecules and macromolecules have been observed at the cellular level; we will generalize these behaviors. In Chapter 8, we introduced a hierarchical description of these media based on biochemistry and ending with eukaryotic cells. We will continue this review by extending it to organs and individuals, mainly in mammals and humans, by introducing the notion of physiology. For this, we will introduce the concept of the internal clock, which regulates the life of animals and man [KIN 14]. Then, the extension to ecological systems is a natural progression. In doing so, we will show the behavioral analogies encountered in populations of individuals but also the difficulty of understanding, in thermodynamic terms, these more collective dynamic phenomena in living units such as ecosystems or in the Earth's natural cycles.

In this approach, in addition to the oscillatory temporal aspect, which is extended to the development of a chaotic situation, we also describe the appearance of spatiotemporal patterns. Thus, within the framework of the Turing model, recent advances in the formation of spatial biological structures will be discussed.

9.1. Behavior of cells and organs

For this, the analysis must evolve from the general definition based on the reservoir–system–observer trilogy (see Figure 2.1). By modeling them together, it is necessary to take into account the content of an open system, its hierarchy in subsystems, but also the concrete definition of the boundaries or selective walls that control the nature of exchanges with the environment. Considering the essential characteristics of a thermodynamic system out of equilibrium, it is also necessary to examine the information contained and the associated cognitive behavior. We will examine this at several levels, from groups of cells, to organs then to individuals and finally to groups of individuals.

9.1.1. *Biochemical oscillations and biological rhythms*

First, we will discuss at the molecular and cellular levels, the bases of dynamic behaviors in living systems. Thus, in a cell, catalytic enzymes (see the allosteric model in biology) form a negative feedback loop and generate a non-stationary response. This is the behavior of a system out of equilibrium that can give rise to a chemical oscillator, such as glycolysis, which conditions a biological cycle [HES 71]. An example of an oscillation is the basic glycogen cycle shown in Figure 9.1 for a yeast, a unicellular fungus. In an open reactor, the cycle shows a periodicity of a few minutes that is dependent on the quantity or flow of fructose injected. This experiment makes it possible to show that the decrease in the injection rate reversibly modulates the biochemical rhythm and allows it, even below a threshold, to disappear to regain a steady state.

Studying a dynamic network of self-catalytic chemical reactions of biological nature is still valid. Indeed, recent work [SEM 16] on reagents relating to cell metabolism or those assumed to play a role in prebiotic systems has been published. In particular, these experiments, conducted in a continuously mixed open reactor, show that a sustained oscillatory regime can be switched to a bistable state. The aim was to establish a link between the competitive onset of temporal or spatial instabilities and the dynamic characteristics of a living cell.

Different cell rhythms exist; examples include one with a periodic peak concentration of calcium ions or associated with mitosis. Thus, the cycles regulating cell division are controlled by a specific enzyme and the onset of chromosome duplication, the basis of cellular morphogenesis. More recent examples of cellular or bacterial colonies have also shown oscillatory cooperative behaviors. Thus, the cell clock can be synchronized by the expression of genes in a bacterial colony [DAN 10]. Indeed, it has been shown that cells communicate with each other by means of a chemical messenger, which is a small molecule (acyl-homoserine lactone) that diffuses through cell membranes.

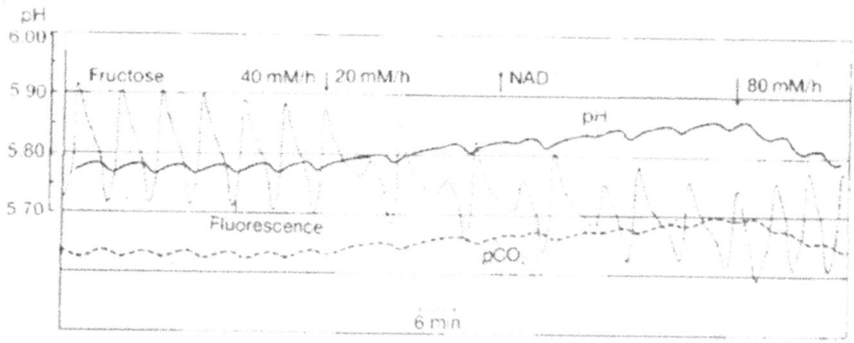

Figure 9.1. *Oscillations of glycogen concentration in a yeast extraction; it possible to show the cycles detected by a fluorescence technique according to the sucrose flow expressed in millimoles per hour; on the Y-axis is the pH of the solution, which is constant, also with the pressure of carbon dioxide present (adapted from [HES 71])*

This communication phenomenon can be generalized to advanced cells and organs even if not all mechanisms are well known. The metabolic basis of these cellular rhythms will be presented in section 9.2 of this chapter. They allow us to interpret the behavior in biological cycles specific to each organ up to the circadian sleep-wake clock in humans and mammals [KIN 14]. These oscillatory phenomena interact with one another, creating a fundamental synchronizing effect for physiological functions and rhythms in living organisms [GOL 97]. They are at the origin of intercellular communication as we have just seen, and play a role in the cell division cycle as well as in cardiac and respiratory organs. These phenomena can go as far as aperiodic oscillations, their dissociation and the onset of a chaotic situation [GOL 12]. Thus, the synchronization of these rhythms and their control appear essential to life. Desynchronization is detrimental to health; their disruption in major organs can lead to certain diseases [KIN 14].

9.1.2. *Spatiotemporal organizations and Turing structures*

Far from equilibrium, a new spatial order resulting from self-organization can appear: it has already been described during the morphogenesis of a material. The Turing molecular model based on reaction–diffusion competition that accounts for this loss of symmetry has been generalized. For this purpose, cell activity must be located in an open system beyond a bifurcation point with the presence of bistable states as shown in the existence diagram introduced previously (see Figure 2.7). Note that the analysis of spatiotemporal structures in living environments has already been introduced in Chapter 2 (see Figure 2.9).

Cooperative self-organization in microorganisms is indeed the cause of the appearance of various motifs [BEN 00]; in a living system, it may be essential during the process of differentiation occurring in multicellular eukaryotes. The generation of patterns by coupling between short-range local activation and long-range inhibition can give rise to a symmetry break [KON 10]. This double activating and inhibitory action in a concentration gradient of morphogenic species underlies this phenomenon in living beings. Meinhardt models [MEI 82] have been used to understand the emergence of these spatial structures in biology, where, as we have seen, the autocatalysis produced by an enzyme provides a nonlinear response. This author has shown that genetic information is responsible for the formation of repetitive patterns and spatially organized regions. Similar models, such as the French flag model by Wolpert [KON 10], have been previously proposed in the context of cell differentiation, where the importance of positional information is introduced. Indeed, the existence of a Turing structure in biological life has been recently proven. It has been experimentally demonstrated that the differential diffusivity of inhibitor and activator proteins is the major factor in the creation of a patterning in living systems [MUL 12]. Thus, a specific interaction between the various compartments appears crucial for the creation of elaborate spatial patterns in various living systems such as in leaves, shells or animal furs.

In embryology, the first stage in the development of an egg, several examples have been published showing that a reaction–diffusion competition model is no longer sufficient. Thus, recently, in these multicellular systems, the influence of gene expression has been shown by the *in vivo* pattern formation in the palate of a mouse or the formation of the finger skeleton in an embryo [RAS 14]. This latter experiment (see Figure 9.2) is interpreted as a Turing mechanism controlling the formation of fingers in a vertebrate. It results from a combination of oriented cell growth, morphogenic gradients and a mechanism of self-organization involving specific proteins. This example shows the delicate transition from chemical morphogenesis, introduced in Chapter 2, to embryogenesis, a key stage in developmental biology.

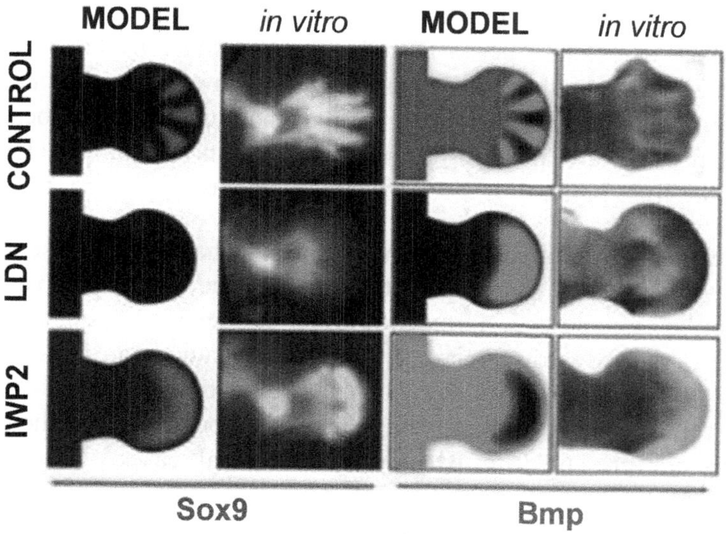

Figure 9.2. *The mechanism of finger development in a mouse paw, showing the respective influences of two transcription proteins of a gene (SOX9 or Bmp). A comparison between a 2D model of the Turing mechanism with experimental observations in vitro and in the presence of inhibitors (e.g. called IWP2 and LDN) confirms this result (adapted from [RAS 14])*

NOTE.– (On the segregation phenomenon in an active system where the particles can move spontaneously under the influence of an internal stress). In a fluid medium, phase separation induced by motility can occur by generating islands with a higher concentration (coarsening phenomenon) [GON 15]. This effect is interpreted using the Cahn Hilliard equations, which minimize the free energy of the system by taking into account the surface energy, as introduced in Box 2.1 of Chapter 2. It is applied in particular to account for the evolutions of bacterial colonies. However, the crucial point in some experimental situations is the difficulty when differentiating between motility and competition far from equilibrium between reaction and diffusion processes.

9.1.3. *Rhythms and chaos in certain organs*

We will now review rhythms and chaos by considering the behavior of certain organs in evolved species by introducing the concept of biological chaos. Among the important oscillatory phenomena, it is necessary to recall the presence of calcium peaks in cells whose periods are controlled by a gene, but also the fact that some organs have their own internal clock. This is the case in the lungs with the breathing frequency or that of heartbeat. These different rhythms are at the origin of the circadian and annual clocks that regulate our lives [GOL 97].

An interesting point relates to the dynamic behaviors of these essential organs such as the heart and the brain. The activity observed in electrocardiology shows a normal operation at a regular frequency but the presence of varied arrhythmias and chaotic behaviors are observed. An example is shown in Figure 9.3 where the electrophysiological signals of cardiac cells *in vitro*, through the temporal dependence of their action potential, are presented in two different clinical cases. The first case corresponds to normal periodic pulsations and the second to severe arrhythmias [XU 15]. Indeed, the heart is a complicated electrical organ whose pathologies can go as far as triggering a local electrical discharge that spreads throughout the heart tissue resembling a chaotic situation. Severe ventricular arrhythmias can cause sudden death.

Neuronal activity of the cervical cortex also appears to be in a deterministic chaotic state as discussed in Chapter 2. These operations are based on electrical pulses that vary in time and space, which give rise to a dynamic state far from thermodynamic equilibrium. Chaotic behavior was demonstrated by the analysis of encephalography signals, indicating that even in deep sleep a chaotic structure may be present; it is different from that recorded in a standby state. The organization of neurons in an unfrozen operational network, associated with brain plasticity, allows for learning in response to specific stimuli distinct from background noise. In neurobiology, the analysis of certain pathologies such as epilepsy results from this [BER 09]. These networks can be associated with particular electrical circuits in a vast set of biological neurons and models have been developed

to understand their functioning. The Hodgkin–Huxley model (Nobel Prize in Medicine of 1963) explains the behavior of a giant squid axon, which is a nonlinear ionic circuit. It describes how the action potentials of neurons, after being chronologically initiated and propagated, can lead to a deterministic chaotic state [GOG 02]. This is a biological version of Chua's electrical model presented in Chapter 3. Then, the behavior of these deterministic systems and their coupling is explained by Kuramoto's mathematical model. In this model, a set of coupled oscillators, with limit cycles, can lead to a synchronized solution under a strong coupling. This neuronal synchronization appears to be fundamental in the cortex functioning
[BRE 10].

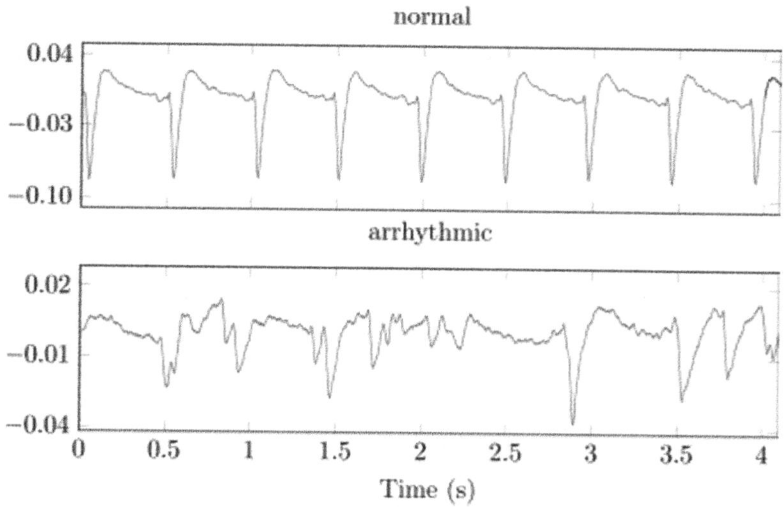

Figure 9.3. *Cellular potentials of in vitro cultures of normal cardiac cells and with severe arrhythmias; the value of the depolarization potential (expressed in mini-volts) is recorded with minielectrodes as a function of time (in seconds) (from [XU 15])*

9.1.4. *Neural networks, information and cognitive behavior*

We have already introduced the concept of sensory detection and the specific role of neurons. In fact, these excitable cells respond to an action potential and together they form an electrical circuit called a neural network. Analogies between the functioning of the human brain and artificial neural networks (see Box 9.1) allow us to model and understand the onset of this chaotic state [KER 03]. It is a set of entities or nodes that are connected according to a certain topology ranging from strictly determined to completely random. There are different types of discrete formal networks that establish electrical connectivity to process and transmit a piece of information: examples are provided in Box 9.1. The first case is a review of simple percolation models [DEG 76] that we have already discussed with regard to transport properties for heterogeneous media (see Figure 9.4(a)). The second case is more complex concerning neural networks with different architectures. The introduction of a more sophisticated topology with more diversified links (Figure 9.4(b)) leads to situations identical to those observed in electrical circuits made up of specific components (see Figure 3.5). Far from equilibrium, time-varying voltage oscillations can generate chaotic behaviors [BRE 10]. In particular, the concept of synaptic plasticity with a modification of the responses, which can be pluralistic on a neuron and summarized according to a variable external stimulus, makes it possible to introduce the concept of learning and acquiring some information. In particular, chronobiology and the frequency of electrical signals transmitted between neurons by synapses are essential to their running. Neurobiology research has helped us to understand cerebral communication and memorization that lead to the storage and addressing of information in the brain. The construction of neuromorphic circuits has enabled the creation of machines capable of learning and reasoning. Artificial intelligence, in particular with regards to the ability to recognize forms and classify them, is created from the detection of senses such as optical analyses of shapes and tactile sensitivity leading to the creation of robots [NOR 14].

The contribution of neuroscience to the cognitive sciences by explaining the mechanisms of knowledge and human thought is a decisive step in psychology. The theory that the brain performs probabilistic reasoning using Bayesian inference to analyze a situation is valid [DEH 14]. This approach is based on Bayle's rule of inverse reasoning to trace the hidden causes in a series of observations and information storage, taking into account the knowledge acquired on the explored system (see the problem of Monty Hall). This is a subjective point of view as opposed to a classical statistical approach, deemed objective, as in a percolation model. This idea was initially proposed by Jaynes, who led to the formulation of a principle of maximum entropy production in statistical physics (see Chapter 5).

The design of discrete models where topology plays an essential role is a branch of graph theory in mathematics. It is a modeling of a system consisting of sites randomly or regularly distributed and a variable number of links between these sites in order to study the transport or communication properties. Here, we deal with two particular cases: percolation theory and artificial neural networks. In addition to the examples given, there are other topologies that are more elaborate, for example, hierarchical (trees) or interactive layers. Sociological networks are the latest virtual representation of these discrete communication systems. Note that these artificial models operate with an electronic current, which is not the case in living systems based on redox processes:

– Percolation models: A regular set of identical sites connected by a constant number of links; they communicate randomly with their neighbors. Depending on the proportion of active links, greater or less than a threshold value, there is a possibility of transmitting long-distance information (see Figure 9.4(a)). It is a critical phenomenon that resembles a thermodynamic phase transition [DEG 76]. A theory based on statistical physics was introduced as early as 1957 by Broadbent and Hammersley and subsequently improved to account for transport properties in a biphasic heterogeneous medium, as an electrically conducting phase distributed in an insulating one. The essential concept is the existence of a threshold with a critical behavior associated with the associated appearance of an infinite cluster of sites or links: an electric current can then pass completely through a sample.

– Artificial neural networks: Formal neural network architectures have been proposed to simulate a set of biological neurons. In this case, the graph is more complicated with different types of sites, each representing a specific neuron with active or inhibitory synaptic bonds. A simple example is given in Figure 9.4(b) to describe the electrical network behavior equated to a set of neurons (sites) and synapses (the links). This conceptual framework has been extended to problems of physics (e.g. in magnetism concerning spin glasses [MEZ 90]) or automata networks, but also for the description of physiological and psychological properties concerning in particular associative memory [NOR 14].

– Communication networks: These dynamic models have been extended to organized networks transferring energy or information: fixed or mobile telephony, distribution and stable electrical "smart grids" or Internet with social networks for example. A global matrix network can process, classify, analyze and retrieve information on demand [ERM 15]. A derived application is to account for the spread of epidemics taking into account the contagion processes. It is the advent of so-called complex networks.

Box 9.1. *Artificial networks and communication*

a) Percolative network model

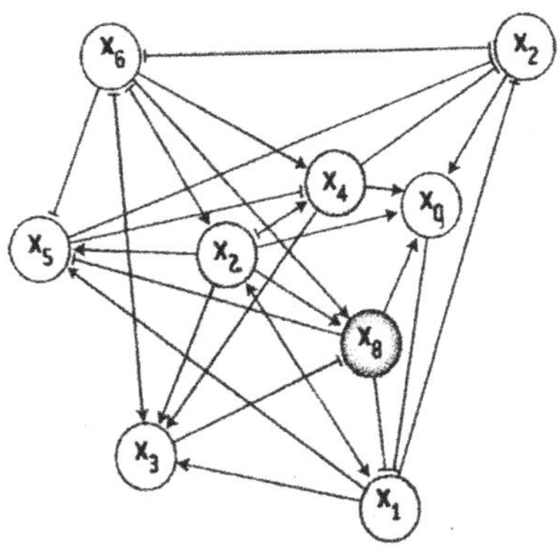

b) Neural network model

Figures 9.4. *(a) Percolation model of links between two sites on a two-dimensional square lattice showing the presence of localized clusters (hatched parts) but also the absence of an infinite size cluster crossing the network to ensure the transmission of information [DEG 76]. (b) Example of a network of formal neurons: nine different neurons (X_i) are represented by circles and the synaptic connections by arrows; they are either excitatory and terminated by an arrowhead, or inhibitory and terminated by a stroke. The analysis as an electrical model of the system can show a periodic, quasi-periodic or even chaotic oscillatory behavior as a function of the imposed external stimuli (according to [BER 09])*

9.2. Physiology of a living organism

We shall examine the physiological characteristics of living entities, that is to say, the functions and properties of multicellular assemblies, organs and tissues. It is a hierarchical level change based on the global analysis of organisms, their critical metabolism for growth, reproduction and aging. The role of the circadian clock is discussed with the importance of the synchronization of the cycles. Finally, the collective behavior of populations of organisms and individuals is analyzed.

9.2.1. *Thermodynamic system and metabolism*

We will continue the thermodynamic analysis of cellular metabolism (see Figure 8.2) of an individual by essentially changing the level of description. To do this, we must start with all the mechanisms and processes that participate in the nutrition and growth of organs and tissues, known as trophicity. As we have introduced, the energy pyramid in animals starts from the production of plants by photosynthesis (see Figure 8.3) with three levels of consumers in animals that are herbivorous, carnivorous or mixed, also detritivorous. Moving from one level to the next, the energy yield is only about 10%, which causes a large loss in each stratum of species; thus, man uses about 0.1% of the natural resources harvested for food.

We shall review mammals by examining the energy scheme presented in Figure 9.5. From a thermodynamic point of view, there are several key points as follows:

– Intermittent open system: In fact, there is an exchange of matter and energy but not continuously as in an open chemical reactor; in general, nutrient intakes occur outside periods of activity and energy consumption. The living system produces energy reserves (anabolism), which it then uses in small quantities (catabolism) according to demand. It is a discontinuous reactor operation that is generally observed having feeding-consumption sequences with several operating modes.

– Functioning of metabolism: In fact, it is necessary to distinguish basal metabolism from metabolism during activity. The first allows the maintenance of the functions inherent to survival and ultimately produces heat to ensure thermoregulation. The influence of caloric restriction affects animal species; their longevity is increased but their fertility is diminished.

For low metabolic rates, the organism modifies its hormonal response to protect itself and mitosis no longer takes place [THO 14]. This regime corresponds to the upper part in Figure 9.5.

– In high metabolic conditions, motor and cardiac activities are linked to muscular functioning; the mechanical efficiency of skeletal muscles is about 50%, the other half being dissipated as heat. The overall efficiency as a human machine is only about 30% because the heat released during the hydrolysis of ATP, mainly during catabolism, consumes 60% of the nutrients. Note that this yield is similar to that of generating machines comprising several subsystems to manufacture electricity (see Figure 4.5).

– Role of the circadian clock: Finally, from an energy point of view, the metabolic processes depend on the day–night cycle and the seasons. It regulates the metabolism efficiency associated with different activities, physical or cerebral and nutrient intake that depends on the time of day [BAS 10]. This point is confirmed by the use of drugs whose effectiveness is related to their synchronization with biological cycles, detected by the emission of biological signals using a sensor. An external stimulus needs to be synchronized to be more efficient.

Physiological responses to intense physical effort have been widely studied by measuring the temporal expenditure of energy or developed power. In general, these are cardiac or respiratory *in situ* methods (e.g. VO_2 max measurements) that show that the maximum intensity of the metabolism can reach several times that of the basal metabolism. As with machines, the increase in power is accompanied by a loss of heat and an increase in entropy production (see Chapter 8, section 8.1.2).

Cerebral activity results in the gain of information detected from the sense organs. Their energy conversion and transfer, as well as analysis and storage, takes place via the neural and hormonal systems. This component belongs to the cognitive sciences, in particular the behavior of the electrical system of the brain, which also consumes energy. Analyzing them as formal neural networks (see Box 9.1) allows us to understand these mechanisms and extend them to evolutionary systems with architectures different to that of a sequential computer. It is a neuroscience approach ranging from the individual functioning of neurons to their much more complex overall behavior [BER 09].

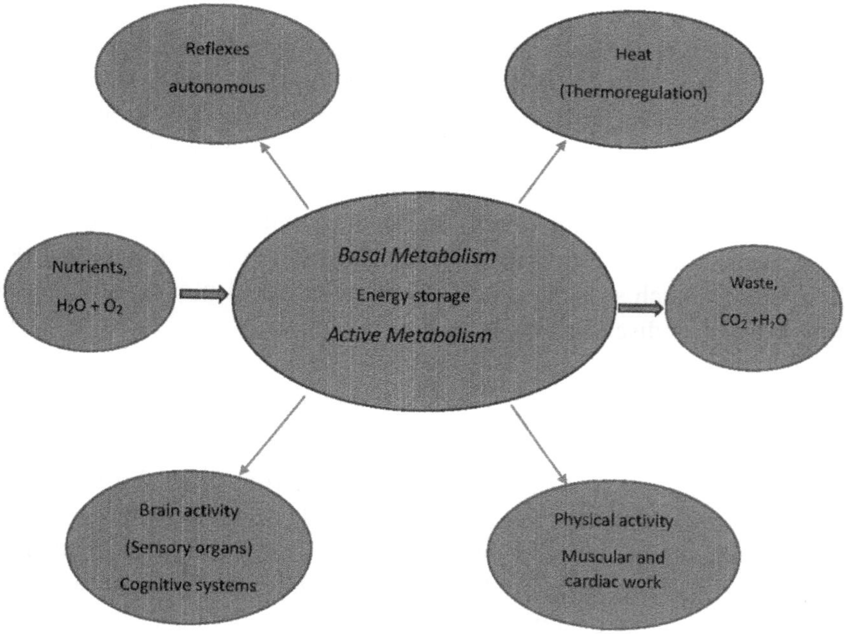

Figure 9.5. *Energy diagram of an open system representing the metabolic activity in a mammal or human*

9.2.2. *Collective behavior*

We are interested in the ability of a living entity to move spontaneously or in reaction to a stimulus in a fluid medium. As we have already indicated, this characteristic is motility, which is the capacity of entities to move spontaneously by consuming the energy supplied by metabolism. Motility-induced phase separation can be observed in fluid active particles such as rod-shaped bacteria [GON 15]. Thus, the emergence of spatial patterns in ecological systems is often explained by this dynamic model, which appears to be the most relevant by inducing a phase separation between states close to equilibrium [LIU 13].

We must consider also the functioning of a group of individuals for a particular species and their overall behavior. Indeed, in a dense bacterial colony, spontaneous large-scale collective oscillations due to a synchronization by exchange information and onset of a long range order are detected [CHE 17]. At a superior level, motions are also present for gregarious species such as a school of fish, a flock of birds or even in the social networks via the Internet. For example, the dynamic behavior of a shoal of fish has been simulated numerically [CAL 14]. It relies on individual interactions that depend on the position and orientation of neighbors. It is a model of cognitive propagation of information that also depends on the importance of the shoal. Thus, a simulation makes it possible to account for different situations in agreement with the observations and show dynamic transitions in an existence diagram as in reaction–diffusion models. Finally, such so-called sociological behavior also exists in humans, but this will not be discussed here.

9.3. Ecosystems and natural cycles

In this context, we will discuss ecosystems starting with the Earth as a closed system, in particular the biosphere, which exchanges only radiative energy coming from the sun but no matter (neglecting the arrival of meteorites or gaseous losses). It ultimately plays the role of a global reservoir of matter for operational subsystems to be defined, which can be open systems subject to a global mass balance. They are present at different sizes in a specific environment: this is the biotope where several living organisms or species that coexist [RAM 03]. A first subsystem may be the microbiota, a set of microorganisms living as parasites within a specific environment. This host environment, which acts as a reservoir of energy and matter, is generally found in a living organism in which the different microorganisms (bacteria, viruses, fungi) are in a cooperation–competition situation, as in the case of intestinal flora or pulmonary microbiota in humans. As such, microbiota can evolve over time and cause chronic pathologies in humans [AND 15]. In this context, it is necessary to take into account the fact that a plant or an animal cannot live without the microorganisms that inhabit it. The organism is extended to the community of these entities in the concept of a holobiont [AND 15]. This is the scaffolding in a thermodynamic system with its environment where all organisms present, i.e. bacteria, viruses and fungi live in symbiosis, which leads us to consider it as an ecosystem with more blurred borders.

On a larger scale, groups of animal and plant populations will constitute classical ecosystems subject to a dynamic environment characterized by variable constraints and flows. The difficulty is then to define the subsystem studied. There are no longer tangible walls that define it explicitly, and the nature and importance of the flows exchanged between subsystems influence the type of response. Their behavior may have spatiotemporal instabilities similar to those described for biological media [HES 71]. For this reason, we will focus on three significant cases: the dynamics of animal populations with prey–predator models, large natural cycles of biogeochemical origin and climatic models.

9.3.1. *The predator–prey relationship*

A particular example in population ecology is that of predators and prey, when they cohabit in a natural zone that is both the system and the reservoir. A common example is that of lynx and hares, the former eating the latter. The variation in abundance of the two species was monitored for a century as shown in Figure 9.6 [RAM 03]. Their population varies with maximum density every 10 years and a small time lag for the predator: this is an example of settlement oscillations in a stable environment.

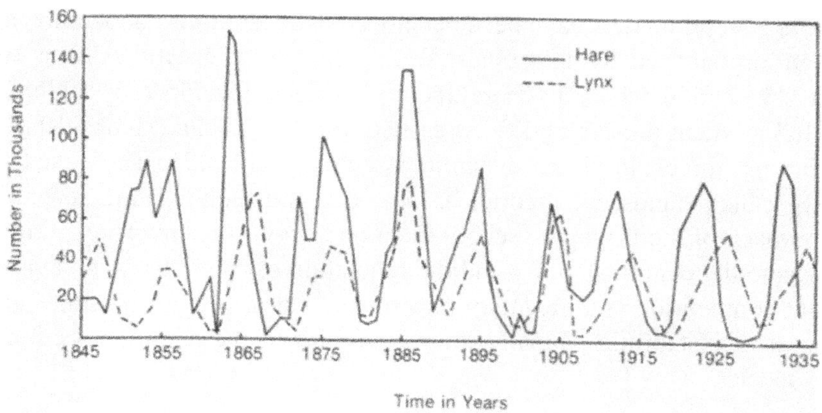

Figure 9.6. *Variations in abundance of variable populations of harés and lynx in northern Canada over a century (according to [RAM 03])*

This model has been validated at the molecular level with different DNA fragments and in the presence in a homogeneous solution of an enzyme playing the roles of prey, predator and nutrients (grass), respectively [PAD 13]. In this prebiotic system, periodic oscillations appear with the visualization by fluorescence of chemical waves. In the laboratory, an open reactor study on the interaction between two bacteria simulating microbiota gave similar results. For example, in a biofilm containing bacteria, there is competition between preservation and deprivation by controlling nutrient intake [LIU 15]. Population expansion occurs through a growth–death process and the appearance of collective population density oscillations related to the metabolism present. More generally, nutrient depletion and reduced metabolism cause the transition to a non-oscillating state confirming the general character of these observations, which are independent to the observation scale.

These population oscillations are interpreted by a chemical model, the ideal Lotka–Volterra model, a set of second-order differential equations that describe a three-step behavior [RAM 03]. Initially, the prey reproduces quickly when its food (grass) is abundant, then the predator who lives at its expense, proliferates and reproduces and subsequently reduces or depletes it. Finally, the predator, if it remains alone, disappears by natural causes. It is a principle of regulation in population dynamics that is proposed for the growth and limitation of living species.

More general models have been developed to account for cooperation and competition between living species. With regard to stochastic systems (see Chapter 5, section 5.1.4), a general theory has been developed by taking the potential gradient present and the presence of different fluxes considered as the driving forces in these dynamic systems [ZHA 12]. These statistic models (called "landscape theories"), located outside equilibrium, generalize the approach of the Prigogine school in an open environment. They allow us to evaluate the overall stability and the regulation of ecosystems, taking into account competition and predation phenomena but also their dynamics until the onset of instabilities. Finally, they form part of the natural selection of living species.

NOTE.– The spread of epidemics with a threshold effect presenting a periodic character is a topic derived from the prey–predator cycle. The spread of human-transmissible infections with different transmission vectors [PAS 14] can be represented by discrete models of neural networks in trees or stars (see Box 9.1). Group dynamics and the propagation of true or false information via the Internet [ERM 15] are also associated sociological topics.

9.3.2. *Grand natural cycles*

Of all the major cycles of matter, that of carbon is the most important. It is based on the storage of carbon of biological origin underground, acting as an energetical reservoir, its transfer into the atmosphere and its assimilation in living media [RAM 03]. In the air, we find it in the form of carbon dioxide and methane, two greenhouse gases that are mainly caused by human activities. When extracted from fossil reserves, its natural cycle is disturbed and the gases emitted also intervene in the oxygen cycle, which can exhibit a particular behavior. Indeed, free oxygen appeared on our planet 2.4 billion years ago as a result of photosynthetic living organisms, cyanobacteria, which produce it by dissociating water. This was the great oxygenation event, one of the major transitions in the history of the Earth. Since the concentration of oxygen has increased, this shift from a low value (less than 10^{-5} of the current concentration) toward a concentration of more than a thousand times higher has not been achieved on a regular basis. A phenomenon of bistability occurred due to the protective shielding effect of the ozone in the troposphere, causing a nonlinear increase in the lifetime of atmospheric oxygen [GOL 06]. Thus, a chemical model presenting a redox behavior of CO_2, but also in the presence of CH_4, provides evidence for a terrestrial atmosphere far from thermodynamic equilibrium. A resulting hysteresis loop with two branches, high and low concentrations separated by an unstable zone, is shown in Figure 9.7.

This behavioral instability of a natural cycle, linked to solar cycles, is a particular example of variations in stresses and fluxes over time. It shows that these spatial or temporal phenomena can occur at all dimensional and temporal scales. Geological periodicities up to billions of years arise through synchronism with astronomical events present in the universe [PUE 14]. For a much more restricted period, corresponding to the quaternary era, warming cycles of approximately 41,000 years were established, explaining the glacial periods [TZE 17]. On an even smaller scale, the location of the Earth in the solar system is not stable because it was initially chaotic, with an orbit, angular position and precession variable that modulate its climate. Figure 9.8 shows periodic fluctuations in average temperatures over the last two millennia [LOE 07]. We see rapid oscillations with average oscillations below 1°C. A warm period in the Middle Ages is followed by cooling around 1600–1700, which appears to be over today. It continues with a currently even greater increase due to the greenhouse effect.

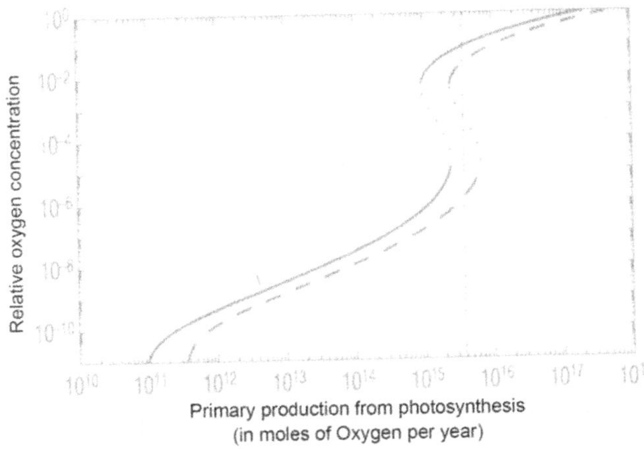

Figure 9.7. *Stationary solution for the redox model giving the atmospheric oxygen concentration, normalized to its present value, according to the annual production by photosynthesis, expressed in logarithmic coordinates: it shows the previous presence of spatial bistability in the atmospheric system (adapted according to [GOL 06])*

9.3.3. *Climate models*

The increasing power of computers has led to the design of elaborated digital programs to simulate the current and future behavior of the biosphere. Several climate models have been developed, taking into account the main elements of atmospheric and oceanic circulations as well as their thermal exchanges. The problem of interactions at different scales and their parametrization are the crucial points addressed in the Intergovernmental Panel on Climate Change (IPCC) evolutionary model [INT 13]. Hydrodynamic movements in the atmosphere and oceans and their coupling can result in nonlinear effects causing turbulence and periodic phenomena (such as possibly the El Niño current in the Pacific Ocean). Without discussing the modeling in detail, we will only cover the global thermodynamic aspect in two points:

– The importance of the energy balance: The Earth receives light radiation and re-emits radiation of a longer wavelength, located in the infrared spectrum. It is subjected to the absorbing effect of water vapor and greenhouse gases, that is to say, with a form of degraded energy. The heat produced through the energy consumption from human activities adds to this phenomenon, which causes a thermal imbalance, heat transfer and global warming. Indeed, for more than 20 years, global warming appears to have been higher than the fluctuations recorded (see Figure 9.8), approximately 1–2°C, which is very rapid compared to these natural variations.

– The production of entropy: An anthropomorphic term causing an additional increase of the entropy is added to the source of natural internal entropy. A release of greenhouse gases into the biosphere leads to an increase (in) related to the mixture entropy. It is a global phenomenon that becomes involved in the entropy balance, expressed in terms of source and flux (see equation [2.8]). If the source term is not counteracted, the thermodynamic system may become unstable. The appearance of a chaotic oscillating state is then possible. Beyond a critical point, which is difficult to assess, climatic instabilities can become more frequent (see the Lorentz weather model introduced in Chapter 2) and cause locally such a phenomenon.

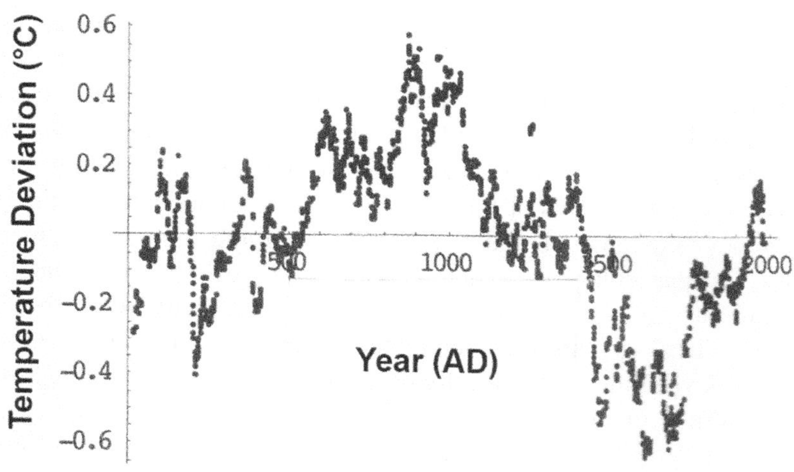

Figure 9.8. *Fluctuations in average temperatures estimated on Earth over the past two millennia (according to Loechle [LOE 07])*

As we have just seen, the climate can change as it is linked, on the one hand, to the variation in solar activity during cycles with eruptions (see Figure 9.8) and, on the other hand, to human activity considering their eventual synchronization. Computer models are becoming increasingly developed with a tight mesh for the exchange of energy and mass between several subsystems. They show a rapid increase in the average temperature associated with the concentration of greenhouse gases and the change in the natural carbon cycle. Under these conditions, a nonlinear response is considered in order to take into account the strong atmospheric–oceanic interactions, which are then governed by the principle of maximal entropy production for the whole complex system [KLE 05].

Finally, to study the evolution of the living world as a whole, we reference the Madingley model, which simulates changes in the biosphere's population on the basis of an interaction with a climate model (www.madingleymodel.org). This numerical model takes into account all the functions and structures of current plant and animal ecosystems to project their future behavior, especially their survival in a neo-Darwinian context.

9.4. Conclusion

We have examined the modeling of a thermodynamic system in living surroundings, multicellular assemblies and vital organs, then in groups of individuals, concluding with ecological systems and natural cycles. They are characterized by increasing complexity under variable study conditions, which are sometimes difficult to define, such as the definition of limits or the existence of selective walls or boundaries. The main issue is the validity of the thermodynamic model and the definition of the associated physical variables, in particular, the energy present in several forms and the expression of an entropy production, associated with the observed evolutions. These systems in a dissipative regime are located at several hierarchical levels, identified in a living being or in ecology or geophysics. They are formed of interactive subsystems, each characterized by specific interactions: physical attraction, chemical messenger and sensitive detection, which transmit the corresponding information. A specific functioning is observed, where the imposed constraints and the recorded flows can vary over time and the concept of an exhaustible reservoir becomes key.

The situations encountered appear comparable to those described in dissipative chemical systems with either a bistable Turing structure or an oscillatory temporal response possibly followed by deterministic chaos. This is the case of the physiological activities of organs such as the brain and the heart, which present a chaotic regime. One of the challenges is to characterize the bifurcation situations from different instabilities up to the onset of a chaotic state. The complexity of responses is closely associated with different rhythms or cycles that have very different sizes and time scales, particularly in geology: this is the key to the hierarchical functioning of the system under consideration. Indeed at the level of our planet, rapid exchanges of matter and energy will cause accelerated climate variations or even instabilities, even if the scenario is not completely defined. Rare natural events are hard to predict, such as tornadoes and tropical cyclones. This must also be considered, as we will see in comparison with the instability of economic models.

10

Application of Thermodynamics to Economy

The economy is gradually becoming an experimental science like physics, chemistry or biology. Our aim here is not to analyze the different economic models, but instead to examine the role that thermodynamics can play in this field, beginning with the fact that controlling energy and materials in all their technological forms is essential. In the case of an economic system, it depends on the action of humans in their natural environment: this is the hypothetical role of an ecological economy. Observed by the chemist Soddy as early as 1926 (Nobel Prize in 1921) and Georgescu-Roegen [GEO 06], wealth is a measure of economic value based on energy. Indeed, thermodynamics have been involved since the advent of the industrial era with the invention of the Watt machine (see Figure 2.2). It is the application of the principles of generalized thermodynamics to a living being, which are then extended to the context of ecology and extrapolated to macroeconomics. We shall first review the main developments proposed in this context and then carry out a critical analysis of the importance of the exchanges or flows involved. We will assess the problem of money in this context and its position in relation to a physical approach, ending with the analysis of out of equilibrium systems in a dissipative regime. Finally, the dynamics of these systems and the existence of cycles with a determined duration, corresponding empirically to the return of the same economic phenomenon, will be approached. We will be faced with a choice of appropriate variables that could define significant energetic and entropic variables. In the final part, we will analyze the influence of information transfers in these increasingly complex dynamic phenomena.

10.1. Thermodynamic models of economy

We will now discuss the main successive energy models that have been developed in recent decades before analyzing the validity of the fundamental concepts involved.

10.1.1. *Chronology of energy models*

Georgescu-Roegen [GEO 06] was the first to show that the creation of an economic science from a mechanistic model is insufficient. He clearly introduced the relevance of a thermodynamic approach of irreversible character (the arrow of time), since the development of human societies cannot be dissociated from the general laws of the physical sciences. The essential question associated with this approach is the extent to which the principles of phenomenological thermodynamics, conservation of energy and entropy increase in an isolated system can be extrapolated to macroeconomics considered as open systems out of equilibrium. This author has made a set of proposals, which have been succinctly summarized [GEO 71]:

– human well-being is a function of the economic production of material goods or wealth;

– it requires usable energy that can be exergy (see its definition in Chapter 2);

– their transformation converts low entropy materials into high entropy materials (typically fossil resources transformed to waste);

– the recycling of high entropy materials is never complete; waste and pollution remain because there is always energy dissipation during the feasible process, corresponding to an increase in entropy (such as, for example, entropy of mixing in a fluid).

Georgescu-Roegen and his advocates analyze this situation finding that low entropy resources are gradually destroyed and depleted even if non-material services linked to information transmission are also taken into account. To do this, it is necessary to optimize and stabilize recycling in order to practice a circular development that can lead to a degrowth, in accordance with the capacity of the stocks available on Earth. Several criticisms have been raised concerning the consistency of the thermodynamic bases used, in particular the concept of entropy growth associated with the evolution of a defined economic system.

The next step is to adapt the economy to thermodynamics in a more quantitative way, as Ayres' work shows, by developing different exchange models [AYR 84]. The basic principle is that the production of goods and services is essentially the result of the flows of matter and energy into and out in a given system. This is the main contribution in the model shown in Figure 10.1, which we will discuss later. When identifying the principles of phenomenological thermodynamics, this author develops several points by considering services as a quantity of information to be conveyed, by defining the concept of efficiency based on the yields of thermal machines (see Chapter 4). The increase in productivity results from increasing energy efficiency, in which energy and raw materials are converted into useful work or exergy. These considerations with regard to technical development lead to the metaphorical idea of industrial metabolism, which must be optimized.

A non-equilibrium thermodynamic approach is developed quantitatively in Kummel's most recent work [KUM 11]. After reviewing the concepts of energy transformation and entropy production in open systems close to equilibrium (see equations [2.5] and [2.8]), he develops an economic system based on labor, capital and creativity to which must be added the necessary energy and raw materials, which produce goods and services (see Figure 10.1). The response of the system in a supposedly stationary state for constraints and stable fluxes is modeled mathematically with "elastic" responses, i.e. a linear first approximation for the resulting global output function. Kummel suggests that wealth comes from economic markets and then establishes a second economic law, indicating that the conversion of energy and the production of entropy determine this global creation of wealth. However, this author does not involve the principle of maximum production of entropy, which controls the evolution of a system far from equilibrium.

As shown in Figure 10.1 for a conventional economic system, the different inputs and outputs of the model show that there is not necessarily a physical unity of variables that appear interdependent and of a different nature. Note that economic flows are defined as movements of goods and services between two entities, which are respectively assimilated to matter energy and information. They are replaced by cash flows, i.e. sales and purchases and more generally the supply–demand relationship in an open financial market. As we have seen, there is then an increase in productivity, the relationship between the production of goods and services and the resources used to obtain it; this ratio is expressed in monetary units. The technological improvements lead to a better energy efficiency of the resource used, and thus a reduction of heat loss and waste. This, accompanied by an increase in trade, is sometimes called the Jevons paradox in economics. An increase of consumption may correspond to the autocatalytic effect observed in nonlinear coupled chemical reactions.

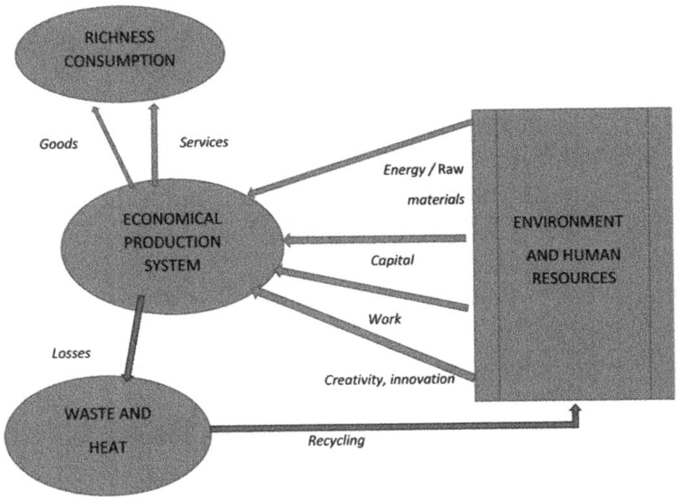

Figure 10.1. *Simplified diagram of an economic system from a conventional reservoir system productive consumption model indicating the main sources necessary for the production of goods and services (adapted according to [AYR 84])*

Thus, identification with a thermodynamic system far from equilibrium [KUM 11] poses some questions:

– the production function is assimilated to a potential function, assuming a system defined by local thermodynamic variables. The variation in this function is studied by a linear perturbation method, analogous to that used for experimental plans. It allows the impact of different variables to be analyzed whatever their physical nature;

– for a strictly thermodynamic approach, it would be necessary to homogenize the flows of inputs as defined physical variables; for example, to be able to transform capital into potential energy, work in kinetic mechanical energy, and creativity or innovation into quantifiable information;

– finally, the definition of an economic system with its tangible limits should be clarified. This is only the case in a few specific examples, such as a company with a specific production plant or a country with political and economic borders, serving as walls or barriers as an input and output filter (see Box 10.1).

The overview of these qualitative approaches is to show that the increase in productivity and growth results from an increased thermodynamic efficiency, accompanied by an increase in economic and monetary flows with entropic consequences. There are also undefined regimes when the exchanged flows are above a bifurcation type threshold as we will see.

With regard to the thermodynamic systems presented in Chapter 2, we can classify the different operation types for a geographical entity, formally defined as a country. The correspondence used is as follows: energy for primary resources, including goods and consumer products created and exchanged, and matter corresponding to population transfers. The economic and associated financial systems operate a priori on a global scale.

It is a simplified approach that allows the following three possibilities:

1) Isolated system: No exchange of product or population: it is self-sufficient. The country must live and develop with its own individuals and wealth. The currency is then only for internal use.

2) Closed system: There is an exchange of resources, consumer goods and services, but not of people, which assumes a selective border. Demography is essentially regulated by the birth rate and trade accounts for the financial equilibrium of imports and exports.

3) Open system: Transfers of goods and individuals are permitted; this situation is dynamic as in an open chemical reactor (see Chapter 2, section 2.2.3). It can lead to an unstable situation depending on the intensities of flows exchanged relative to current stocks. This can lead either to oscillations of economic production and trade, or to spatial segregation of wealth or populations.

We must include two concepts to this approach: limits or boundaries and structural hierarchy. The borders of a country play a selective role as does the membrane of a living cell (see Figure 8.1), which ensures the following functions: to contain, protect, guarantee identity and filter trade. They depend on the political model and the laws in force (customs duties, authorization of entry, etc.).

Increasing political levels establishing successive strata will further complicate the problem. Take the European Union as an example where three levels exist. The first local or regional level is defined as that where social-emotional exchanges are present: in the broad sense, it is the community of classical life (family, tribe, work and relationships, etc.). Then, the second national level corresponds to the country entity and covers the current concept of nation state. This level is essentially governed by political behavior based on a legal set defining a sovereign state. Finally, the supranational level, European in this case, is based on the concept of civilization, a concept reflecting the state of a society with common historical and sociological values. It is more global than the concepts of state and people, which are often associated.

These three levels correspond to closed or open systems that operate in a complex unit with different subsystems interacting with one another. This hierarchy, which must be logical, can be compared with that of a living entity, as we have already presented (see Chapter 8) with the concept of functional complementarity (or subsidiarity) and its own rhythms. This qualitative description shows the need for internal coherence, with the highest level normally controlling the entry and exit of goods and people, even though an internal partial selectivity may exist. In this context, the free movement of goods with a single currency (i.e. existence of a Euro zone as a subsystem) and possibly that of persons within (i.e. Schengen area) can be established ideally inside the same zone according to common rules. The comparison can be made with a living system: differentiated eukaryotic cells (the regions) forming an organ or an individual (the country or state) and then a group of individuals or populations acting together (supranational group). The global level then corresponds to an ecosystem (or macrobiota) subjected to cooperation-competition between states or federations acting as populations of diverse given species.

This has begun to be exploited using dynamic models, such as a formal neural network, which takes into account the various exchanges with nonlinear phenomena. The bifurcation thresholds, i.e. instabilities, can be modeled in the presence of increasing trade flows and population transfers. For example, a network analysis of European countries shows that the most connected economies are the most vulnerable ones. In a closed system, the propagation of an economic shock can only be controlled by political intervention, influencing trade in goods and services [ALA 14].

Therefore, a need for internal coherence, as living systems, is indispensable for the proper functioning of these organizations, as long as they behave as thermodynamics systems.

Box 10.1. *Social–political systems*

10.1.2. *Analysis of fundamental concepts*

The main question is whether an economic system can be also defined in thermodynamic terms. This question was approached in particular by Passet, under the term developmental thermodynamics, by assigning it a limited but fundamental place [PAS 96]. In particular, this author insisted on the role of exchanges in an open system, which is not in a state close to equilibrium, and therefore in a nonlinear response situation near a bifurcation threshold. Passet considers that an economic system is a dissipative structure and that a sufficient energy supply can cause self-organization as in a living system. An intrinsic complication is introduced when incorporating additional information into the energy flow supplied to the considered system.

10.1.2.1. *Note on the role of currency*

As we have already indicated, currency is a crucial parameter. Since its introduction by Aristotle, currency is an instrument of artificial payment for exchanges between the buyer and seller. It is also a supposedly fixed store of value and a generalized unit of account for a set of transactions. This currency, which at the beginning was an intermediate means to facilitate trade, has gradually become a dematerialized entity, created by state banks. The quantity of money, or monetary mass in circulation, then becomes an expanding parameter, which is difficult to control. This evolution, knowing that there are many currencies with variable parities, has led to the emergence of interactive financial systems, each with its own existence: they must then be distinguished from intrinsic economic systems. A financial system is then considered as a peculiar closed system, characterized by independent money flows in the absence of any transformation of matter, although it should serve as a reservoir (capital) for an economic system.

Currencies allow you to convert, store and transport wealth. This definition can be compared to that of energy and its many facets because they account for the production, distribution, exchange and consumption of goods and services in a society. The quantity of money is not conserved as energy in a dissipative system [RUT 05a]. Mathematical models based on different assumptions have been proposed to account for economic developments and associated cash flows. There are several economic schools, liberal and monetarist, Keynesian or Marxist schools, etc. This does not concern us here insofar as it is not possible to consider an ideal, unique and stable currency, giving a reproducible cost of energy. By behaving like a physical variable, the economic activity could then be ideally expressed as the product of this extensive variable associated with another intensive variable in this case an economic potential, analogous to a chemical potential (see Table 2.1).

10.1.2.2. *Validity of the concept of system and subsystem*

As we saw in Chapter 9, the Earth is a closed system that exchanges energy with the Sun, an essential source of radiative energy, which behaves like a reservoir. A specific economic unit, such as global car production or a particular brand in a given country, will be an open subsystem that exchanges matter and energy. The internal transformation into a finished object such as a car is analogous to that of a chemical reactor, operating with a given efficiency characterized by the constraints and flows or exchanges present. The constraints are imposed and within the context of a political–economic system: these are regulations and customs duties regulating or even selecting international trade of import–export, permitted within the framework of a global market (see Box 10.1). The incoming and outgoing flows are, on the one hand, the energy and raw materials required and, on the other hand, cars and waste (see Figure 10.1). Under these constant constraints, we must then assess the variation in production over the years, a problem which we shall discuss in section 10.1.2.3.

With regard only to flows of matter, as we have seen, the general condition of closure of all these subsystems will be such that the total matter exchanged on Earth is zero. It is the role of fossil and mining energy resources, but with the depletion of these stocks or reserves we must rely only on the recycling of certain material goods of inorganic origin and the biosphere production. A circular economy becomes increasingly present, but this is not represented by the diagram in Figure 10.1. The used matter is preferred but it is controlled by the energy required to carry out such recycling. With typical examples, e.g. recycling of glass, scrap or aluminum by melting, the necessary energy is then heat to melt and purify the material. This degradation of energy, which then leads to maximum entropy production, is not optimal in terms of exergy efficiency in this process, favoring the recycling of raw materials. This approach should be more integrated into the overall concept of sustainable development and the assessment of the optimization of the rate of energy return in a controlled system.

10.1.2.3. *The meaning of energy and entropy*

Using the point made in the introduction, there is a link between primary energy consumption and the wealth characterized by the gross domestic product (GDP) in a given country. As shown in Figure 10.2, there is an approximately linear relationship over the years between the primary energy consumed and the wealth of the states by taking the GDP as an imperfect parameter, quantifying this wealth as it is currently calculated.

These studies show that the role of energy in the economy is underestimated; typical cost share models, which indicate an influence of about 10%, are generally exceeded [GIR 17]. It was the impact of energy consumption on GDP that was the main phenomenon, which had already been proposed by Ayres and Kummel (see section 10.1.1). There is an empirical relationship between the cause, the level of primary energy consumption and the effect, the growth rate. Indeed, this relationship (Figure 10.2) indicates that the dominant factor is energy growth, but the efficiency of use and conversion must also be taken into account (explained by Kaha's relation). Different forms of energy are operational and converted with variable thermodynamic yield. These different forms are presented in Figure 2.3, highlighting the particular role of electricity, which is the most frequently used secondary energy. These energy yields assessed from heat (Carnot's yield) or other forms of primary energy are crucial (see Table 4.2). They can be extended to economic systems where the analysis of energy balances becomes more delicate and the associated productivity increase does not seem to have been quantitatively developed. Better energy efficiency must result in a lower consumption of equivalent GDP and an inflection to the left of the drawn straight line in Figure 10.2: this is what is beginning to be observed for some European economies [JAN 17]. In summary, a deviation from this empirical linear relationship is an indication of the beginning of an energy transition. The key to this transition is the improvement of sustainable or biomimetic energy conversions with increased storage possibilities (as described in Chapter 4).

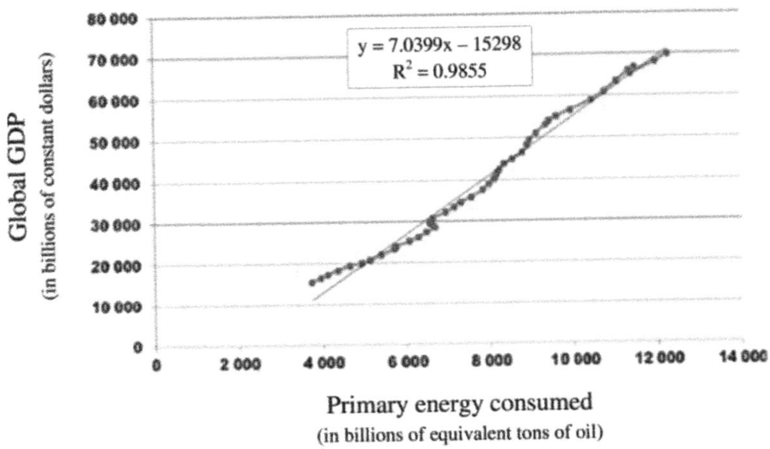

Figure 10.2. *Linear relationship (see parameterization equation and its reliability factor R) between the estimated increase in world wealth expressed as an overall GDP and the growth of primary energy consumed in half a century (1960–2010) (adapted from [JAN 17] and [GIR 17])*

Moreover, the actual conversions are irreversible in nature and produce entropy. Indeed, in an economic system, the optimal efficiency of a quasi-reversible transformation is sacrificed in favor of the energy power, translated in terms of productivity. Indeed, in the stationary state, the internal variation of entropy must be equal in absolute value to that of the flows exchanged (see equation [2.7]). Its increase is then linked to the irreversible nature of the transformations, which indicates the arrow of time. This is the meaning of the second law given by Kummel [KUM 11], which indicates that the rates of energy conversions and associated entropy production determine the growth of wealth. Operational systems are often far from equilibrium and we must rely on the principle of maximal production of entropy, which we introduced at the end of Chapter 5. Under these conditions, one should consider a variable equivalent to the classical entropy, sometimes called economic entropy. Such a generalization of the concept of entropy [JAY 91] introduced as criterion of evolution is still something to overcome.

10.2. Dynamics of economic and financial systems

Economic systems are characterized by fluctuations in production, observed in economics as changes in GDP at the national or global levels. These are periods of growth and decline that result in the reversal of variations in economic and financial parameters. These cycles of short or long durations are analogous to the biological and geological rhythms that we presented in Chapter 9. They will be linked to a variation of fluxes associated with specific reservoirs (see Figure 10.1) identified in a situation far from equilibrium. An analysis of the dynamic evolution of systems led Schumpeter to explain the presence of several industrial cycles of varying lengths and to draw attention to the availability of resources [SCH 39]. We will review the main economic cycles in the context of a classical economic analysis [CLE 17].

10.2.1. *Economic cycles*

10.2.1.1. *Short-term or minor cycles*

The Kitchin cycle corresponds to the reaction time in an enterprise in response to a new commercial situation (likened to a stimulus) and translates into a price fluctuation. This cycle, lasting 3–4 years, is linked to the taking of information and a variation in flow associated with work, and therefore with the labor employed. There is another cycle, called the Juglar cycle, which is generally composed of two to three Kitchin cycles, with a period of about 10 years [CLE 17]. This monetary cycle is linked to a change in investment and issued from a capital flow. In these two cycles, we find the two conventional reservoirs, labor and capital, as bases of an open economy by considering the energy parameter as constant.

10.2.1.2. *Kondratiev waves*

Without describing this cycle in detail, it is a cycle of about half a century and is divided into two equal phases. Phase A representing expansion and phase B, the recession of an equivalent duration. According to Schumpeter [SCH 39], the end of this cycle is linked to the creative destruction and the generation of a new series of innovations that start a novel cycle. At the end of the cycle, these technological innovations provoke the beginning of the following phase A of expansion. These clusters of technological innovations are associated with industrial developments that have been described, each bringing new economic growth. These Kondratiev waves can be associated with primary energy production and the nature of the tanks used, fossil or sustainable energies. As we have shown in Figure 10.2, this primary energy is a largely underestimated parameter [GIR 17]. These are the energy transitions that have taken place since the beginning of the industrial era, thus reflecting technological changes. Indeed, over the last few centuries, the main source of energy has changed, approximately every half-century as for the Kondratiev cycles. At the end of the 19th Century, coal replaced biomass, and after the Second World War, oil and natural gas became the predominant fossil resources. Currently, oil seems to have reached its peak and natural gas is expected to reach its peak soon. The peak of coal extraction will occur around 2030–2050 as it is a cheap and well geographically distributed source still with large stocks, even though it is highly polluting. It will mark the end of a cycle caused by the gradual depletion of all these fossil resources (including Uranium), and the increase in extraction cost will make them more difficult [INM 13]. Thus, the appearance of a new long cycle, whose apogee will occur toward the end of the 21st Century, has been declared. This energy transition can only lead toward renewable energies for the end of the century, and is therefore mainly derived from solar radiations (photovoltaic and photothermal ways). It is associated with the idea of a third major industrial revolution [RIF 12].

10.2.2. *Analysis of financial fluctuations*

It seems rather delicate to appreciate the different economic cycles on the basis of macrofinancial analyses. In Figure 10.3, we present an example of the real GDP of the United States for the period 1954–2005, which shows irregular growth linked to oscillating energy consumption [BUI 17]. By plotting the logarithm of the deviations of the average GDP, independent of the observed background noise, periods of growth are observed, varying cyclically. Oscillations of a few years are observed too, which do not exhibit a regular rhythm and only a spectral analysis would make it possible to identify them as short economic cycles.

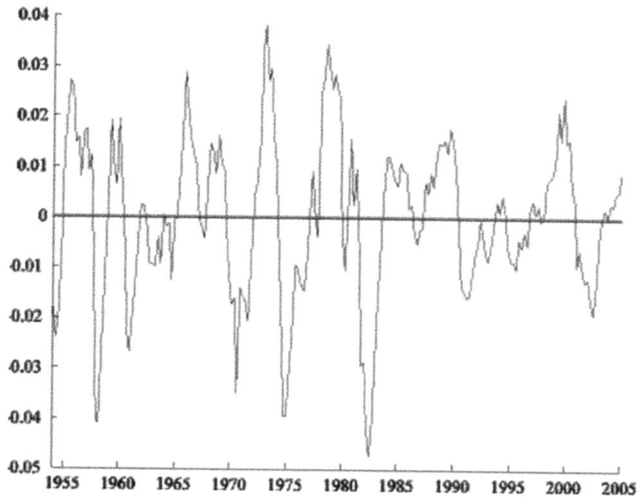

Figure 10.3. *Half-century oscillations of GDP of the United States by plotting the logarithm of the annual deviations versus its overall variation along the Y-axis (according to [BUI 17])*

We see that controlling various parameters involved in an economic-financial system, with cycles of their own periodicity cycles, is a delicate operation. In any case, this and other examples in the literature show the existence of open-minded behavior in dynamic regime far from equilibrium. They are conditioned by different reservoirs that can be interconnected together and present a specific feedback loop that partially conditions the market economy [AYR 84]. They generate oscillations of production with different periods, which can be conjugated or opposed according to their synchronization, which becomes a fundamental parameter. As with living organisms, it is a typical case of complex hierarchical systems, which must have various oscillatory regimes, which can ideally be in phase.

10.2.3. *Stock market crashes*

In this context, economic crises revealed by the major international financial crises (for example in 1929 and 2008) are analyzed as systems with giant financial fluctuations analogous to those of a fluid circulating in a turbulent regime. In Figure 10.4, a standardized variation of the U.S. stock exchange over a century is presented. It is the ratio between the share price and corporate profits in the United States that shows a variation with main peaks obtained in 1929 and after 2000, followed by the sharp decline in stock prices, qualified as major crashes.

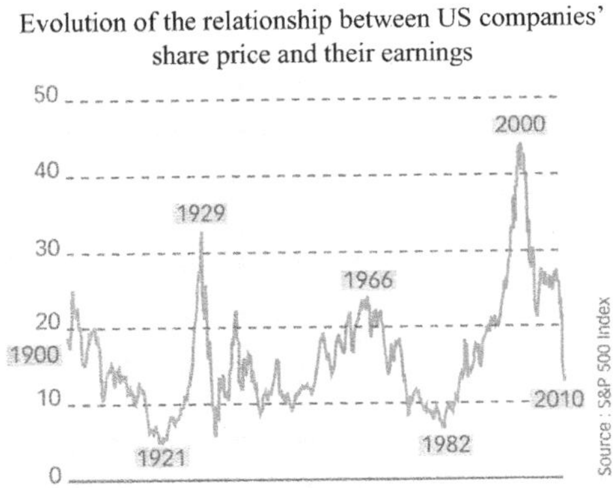

Figure 10.4. *Time evolution of the ratio between share price and profits of American companies, evaluated by the firm Standard and Pools, using the index 500 of the New York Stock Exchange*

As we indicated, currency is a financial parameter introducing an additional variable to the economy (it is like a variable of a variable). However, the volatility of stock markets is an indicator of economic dynamics. In this context, financial models have been proposed in the context of the deterministic chaos theory, introduced in Chapter 2, to establish a dynamic scenario. Thus, in a statistical approach, Guégan [GUE 03] developed nonlinear mathematical models to detect the existence of chaotic behavior from a set of experimental data. These stochastic models analyze the presence or facilitate the suppression of a deterministic chaotic state, considered as a rare event.

10.2.4. *Statistical modeling of financial systems*

From a more general point of view, the question that has arisen is whether the ideas and methodology developed in physics can be used to develop a science of economics. The modeling of stock market fluctuations and financial risks was addressed by examining the distribution tails of these stochastic models [BOU 04]. It is due to the presence of power laws in the distribution tails of an event, or even for temporal correlations of a significant variable, whose statistics obey the laws of Levy (see Chapter 5).

They are characterized by large distribution tails with a power law (Pareto's law in economics) that allow a better approach to market laws [CON 09].

They were originally used by Mandelbrot [MAN 14] in his fractal geometry applied to economics and finance. This author has shown that the turbulence of financial markets can be modeled using the fractal theory, accounting for variations in financial data at different time scales. Using, for example, a mathematical tool as a wavelet analysis of a signal, such as stock market fluctuations, progress has been made in the monitoring of financial markets [ARN 98]. This technique, which is a local Fourier transform, allows for the decomposition of a function in time and space. It was first applied to interpret the growth of materials, such as snow crystals and dendrites (see Figures 2.11 and 2.12) and then extended to detect the reversal of the trend in a financial cycle. An example is shown in Figure 10.5, showing the index of industrial production for the euro zone for over 40 years. Hatched areas are periods of recession that do not appear in a strictly periodic manner, but they can be compared to a set of interferences with short economic cycles. The challenge is to predict when industrial activities and European economies (via GDP) will return to growth [ADD 13].

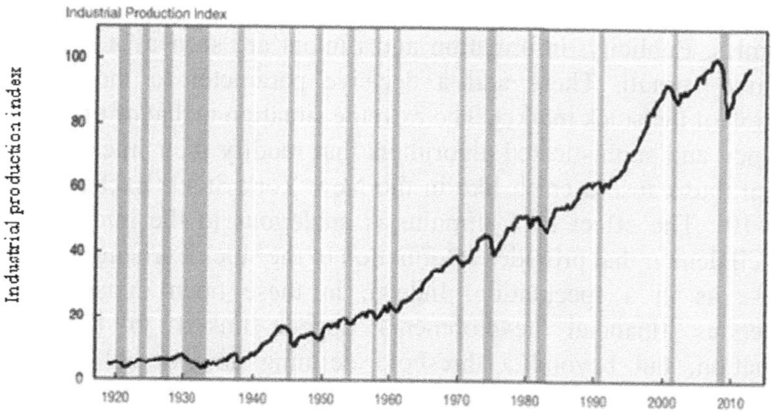

Figure 10.5. *Financial fluctuations in the euro zone, defined as an operating system, showing the growth of the European Union industrial production index with its quasi-periodic phases of growth and recession (shaded areas) (according to [ADD 13])*

Indeed, the concept of random nature is opposed to benign randomness present with noise in a Gaussian behavior. The statistical approaches are based on extreme values account for rare phenomena (or "black swans") such as financial or speculative bubbles. By generalizing the modeling of risks of natural disasters (earthquakes, tsunamis, floods, etc.) or those created by man (nuclear power plant accident), it goes beyond the use of the stable laws of Lévy [CON 09]. For example, the study of complex dynamical systems, involving a network of oscillators in interaction with a chaotic attractor, has been proposed as a behavior model [DES 13].

10.2.5. *On the behavior of a financial system*

In the previous section, we presented this in econometrics frame as an experimental system that could be analyzed in dynamic terms by the importance of incoming and outgoing financial flows. However, for these systems that account for changes in the stock market, it appears that the analysis in terms of flows and stocks is not sufficient, but in fact rests on another variable. In addition to the economic exchange processes, capital flows in an open political system (see Box 10.1); here there is some phychological context. This last point represents the non-rational influence of the financial market stakeholders, who bring either confidence or uncertainty with the management of supposed risks. One branch of research that takes this into account has been proposed, known as behavioral economics. Publicity, information and rumors are seen as a succession of disruptive stimuli. These add a decisive parameter to the spontaneous evolution of the stock market. The extreme situation is that of the use of high frequency and sophisticated algorithms that modify their internal dynamics and can thus cause a crash like in the New York Stock Exchange recently [COL 10]. The effect of a stimulus is analogous to the intervention of a Maxwell demon that provides information to the stock exchange system, true or false as in a speculation. Indeed, in these open dynamic systems, endogenous financial developments appear linked to this flow of information, but beyond a threshold defining an external shock as an algorithm trading can also modify its future behavior. The resulting system dynamics function of its own response time is altered and can induce a new situation, stable or unstable. It is a decision-making mechanism that has a cognitive character that is a Bayesian inference with the strategic use of information. It is analogous to that introduced in a cognitive model developed for artificial intelligence (see Chapter 9, section 9.1.4). This recent approach is extended by works on risk analysis and game theory.

10.3. Conclusion

A thermoeconomic analysis in terms of energy and entropy is only qualitatively addressed in the literature. Historical approaches have made it possible to establish correlations by showing that thermodynamics out of equilibrium is the adapted model in a first approach. Energy as a primary resource, to which should be added the supply of raw materials that will become scarce, plays a critical role as reservoirs, although it is often underestimated. It is also necessary to consider situations not in a stationary state but rather of an oscillating or bistable nature and which can even adopt a chaotic state as in a financial crash. A quantitative approach poses the problem of defining relevant intensive variables as in a physical system and of quantifying associated extensive thermodynamic variables, particularly the generalized concept of entropy. It should be pursued at a strictly energy level to better understand its influence on ecosystems that can explain global warming and the associated climate change.

A statistical physics approach in economics then shows that these financial systems possess internal dynamics with their own rhythms analogous to those of living environments or great natural cycles. Experimental sources show, however, that these periodic characters are not as obvious as for the physical, chemical and biological systems. Stochastic tools based on other distribution laws such as Lévy's laws must be applied. In this context with natural, industrial or financial cataclysms, which are rare and not very predictable, we find financial bubbles and crashes, but also earthquakes or tropical cyclones [GRO 09]. In order to act, it is necessary to establish a strategy to therefore bring significant external information and introduce the concepts of adaptation and resilience in the response of a system that has become complex.

11

From Thermodynamic Systems to Complex Systems

This analysis is based on the methodology used throughout the book. The crucial point is the definition of a thermodynamic model, introduced in Chapter 2 as a system representing the essential characteristics of an apprehended reality. To do this, significant variables interrelated by mathematical relationships are defined. The physical model that is subjected to constraints is analyzed, in the presence of flows and stimuli supplied by reservoirs or stocks, with answers classified from the major concepts of phenomenological thermodynamics. A microscopic description completes this approach by introducing the statistical physics of particles and quantum mechanics as well as the notion of information. An epistemological incursion, based on the comparison of micro–macro approaches, is thus developed from the thermodynamic classification of materials. It is summarized in the second part of this chapter and then extended to living, ecological and economic systems. From inert systems to living environments, various interactions and the symbiosis between man and machine can be generalized and extrapolated. This approach, described by De Rosnay, is part of a global systemic approach [DER 95]. These main points will be discussed in the third part of this chapter with a transversal analysis. It will allow us to expand the concepts of reductionism and complexity by showing the current limits of thermodynamics. The developing concept of complexity then appears: it is context dependent, sometimes ambiguous and needs to be clarified.

11.1. Thermodynamic models: from energy to entropy

11.1.1. *Modeling of a thermodynamic system*

We again start from the definition presented in Figure 2.1 and all the extensions proposed later. We first recalled the thermodynamic definitions in an isolated system and then in a system that exchanges with its environment. The key point that will follow is the application limits of the macroscopic model.

11.1.1.1. *The textbook case for an isolated system*

In this equilibrium situation, where the two principles of phenomenological thermodynamics apply, the state functions are defined as the product of an intensive and extensive variable (see Table 2.1). In this context, the notion of reversible transformation without entropy increase is the ideal situation rarely encountered. Any production of energy or power supplied involves heat degradation of this energy with an increase in entropy. It is therefore obvious that, in a real system, there is a necessary exchange with the exterior, even if only with the existing environment. The nature and flow of these exchanges through an enclosure then becomes important. It is then necessary to extrapolate the exact relations established for non-equilibrium conditions.

11.1.1.2. *Real, closed or open systems*

Two parameters are relevant: on the one hand, the definition and the role of the enclosure or wall between the system and its environment and, on the other hand, the quantity of exchanges and their temporal stability. This last point implies the durability of stock in a given reservoir.

– The enclosure or wall defines the boundary to the exterior: it is characterized by a spatial constraint but also by its specificity. Indeed, it is selective when it passes a form of energy or certain molecules through the interface, as in the membrane of a living cell (see Figure 8.1). Its selectivity of molecules and of electrical charges that can pass through is a characteristic of the functioning of a given living system. In the case of ecological or economic systems, its definition becomes trickier and it is the whole system environment including the presence of reservoirs that must be taken into account. Finally, the existence of specific internal walls leads us to consider subsystems interacting with one another. Indeed, their functionality or specificity is hierarchized in the ensemble considered. This is already the case in a eukaryotic cell, but also in its association to form an organ, an individual or even in a microbiota. This point will be discussed in the second part of this chapter with the concepts of hierarchy and complexity.

– The nature and the importance of flows: these are flows of thermal, electrical or optical energy that may be associated with flows of matter. For low fluxes, a first-order perturbation linear-response of the thermodynamic state functions allows us to analyze the system response within the context of a causality principle linking a specific cause and observed effect. It is essentially based on Onsager's relations for homogeneous materials (see section 2.2.1 in Chapter 2). Starting from a certain level, difficult to evaluate at times, a nonlinear response is characterized by the presence of a bifurcation point; the linear approximation becomes insufficient and a new behavior is observed. A new organization linked to dynamic transition phenomena appears, described by different types of bifurcations and translating the transition from a conservative to a dissipative energy state. The ultimate presence of deterministic chaos is the eventual behavior of these dynamic system.

– Permanence of flows and reservoirs: in thermodynamic terms, these are stationary states defined by an energy balance equation for imposed constraints and supposedly stable flows. An oscillating regime is no longer preserved if the incoming flows decrease, due to the exhaustion of one or more reservoirs. This is the case for an oscillating chemical reaction in a homogeneous unpowered reactor, the reduced intake of nutrients in a living organism or the decrease in fossil energy reserves in the economy. It is also the case for punctual stimuli of an informational nature that may correspond to a transient state, dependent on the system response time, in order to find or reach a new stationary state (see high-frequency trading in stock exchanges).

Generally, an increase in fluxes leads to a change in regime for a system that then has a temporal or spatial structure, a situation corresponding to a dissipative state, which is expected with nonlinear responses in the presence of feedback loops. These possibilities are shown in Figure 11.1 by comparing examples of optical or conductive materials of electricity, living environments and macroeconomic systems. This summary clearly shows a behavioral similarity as a function of the deviation from an equilibrium seenario. Note, however, that there are different ways to obtain a cascade of bifurcations other than a limit cycle and an attraction basin (see section

11.3.4). They are not often explicated, i.e. in the definition of a critical value for which a flow of matter and energy or even information leads to oscillatory instability. A bi- or multistate spatial discrimination described using the Turing model may exist from a competition between reaction and diffusion. Let us also note that the Cahn–Hilliard mechanism (see Box 2.1) also exists in particular situations by giving spatial structures to neighboring shapes. Furthermore, self-organizations could also appear when the supply of matter is affected by a convection mechanism or during the propagation of chemical waves which rearrange the space.

To summarize, dissipative structures exist far from equilibrium in order to optimize the flow that passes through them. They maximize the rate at which energy dissipates in the system as indicated by the principle of maximum entropy production.

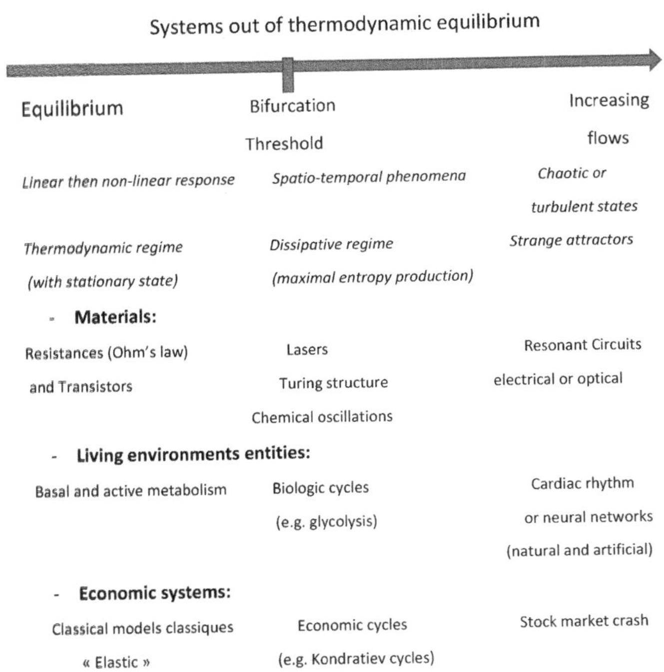

Figure 11.1. *Schematic representation of the evolution of thermodynamic systems situated more or less distant from equilibrium according to increasing energy and/or matter flows and in the presence of feedback loops; examples corresponding to the different regimes are given in the case of materials, living entities and macroeconomic systems*

11.1.2. *Entropy and information*

Energy is a physical quantity that can exist in different forms and for which the conversion or transformation factor must be optimized by reducing heat loss (see Figure 2.3). The concept of entropy, on the other hand, has a more complex meaning linked to its formulation; as we pointed out in Chapter 5, there is a modulation between its phenomenological definition and that identified at the microscopic level. We will discuss the different aspects of this abstract notion.

11.1.2.1. *Phenomenological and statistical evolutions*

As we have seen in Chapter 2, the macroscopic notion of entropy arose from Carnot's work and the formulation given by Clausius. Note that entropy is an extensive thermodynamic variable that characterizes the degradation of one of the forms of energy during a non-reversible transformation or during some use. From a practical point of view, exergetic analysis makes it possible to practically optimize these transformations and minimize the increase in entropy. To summarize this approach we can review a century-old Bergson text. It is extracted from "The Creative Evolution", a philosophical study focused on the meaning of life [BER 07]:

> "It (entropy) essentially expresses that all physical changes have a tendency to degrade into heat, and that the heat itself tends to distribute itself in a uniform manner between objects. In this less precise form, it becomes independent of any convention; it is the most metaphysical of the laws of physics, in that it shows us, without interposed symbols, without measuring instruments, the direction in which the world is moving".

This metaphysical character, characterized by the arrow of time, is prevailing. Various phenomenological entropy formulations have been proposed giving it a different meaning to better define and measure this magnitude [VEL 11]. A decisive step for systems that exchange energy and matter was made with the Prigogine school [GLA 71] based on the entropy balance (see equations [2.6] and [2.7] in Chapter 2). The entropy variation is divided into two terms, the contribution of the external environment related to the entropic flux and the internal entropic production generated from the irreversible processes. This positive source term then induces system stability and a general criterion of evolution that goes beyond the second generalized principle: this is the concept of maximal entropy production in evolutive systems. Its physical meaning, however, is deepened at the microscopic level by looking for the particle trajectories in the phase space (see equation [5.4] in Chapter 5). This is the main contribution of stochastic models for small systems arbitrarily far from an equilibrium situation, which is useful in nanoscience [VEL 11].

11.1.2.2. *Statistical entropy and information*

Indeed, in the historical context of a classical statistical distribution, Boltzmann entropy (see equation [5.1] in Chapter 5) expresses disorder in an isolated system. At equilibrium, all energy states are equally probable: it is a microcanonical ensemble. A generalization has been introduced by Gibbs expressing the entropy from the probability distributions in a thermodynamic system which exchange with the exterior (canonical ensembles). The discrete formulation of this Boltzmann–Gibbs statistical entropy, based on a distribution of the particles following a Gaussian distribution law, is written as:

$$S_{B\text{-}G} = -k_B \sum_i p_i \,\mathrm{Log}\, p_i \qquad [11.1]$$

where p_i is the probability of occupying a microscopic state i.

However, equation [11.1] has limitations corresponding to the need which resulted in the use of other statistical distributions. As we have already introduced, this is the case in certain physical situations and when tackling economic problems. Schematically, this normal Maxwell–Boltzmann distribution function is modified in two main situations:

– either in the presence of natural phenomena, rare or singular events as introduced at the end of Chapter 10. The system presents a non-Gaussian distribution and the Lévy statistic on extreme values is applied;

– or in the presence of additional information brought to the system by the observer. It is external, i.e. a stimuli, or internal as in the case of the genetic code in a living environment. It then forms part of the Bayesian approach starting from the known causes in the system.

We have seen that information has a thermodynamic content that must be taken into account in the exchanges between the system and the external environment. This is particularly the case for feedback loops whose prototype is an isothermal system comprising a thermostat. The nature of the flow circulating in the loop will condition its operation; it can be mass (chemical reactions), energy (heat for example) or information (acquisition or erasure of data).

This approach allowed Jaynes [JAY 03] to generalize the expression of entropy from non-equilibrium statistical mechanics in terms of information theory by introducing the Bayes formulation based on conditional probability. When new information is provided to the system, the probabilities are reestimated. The difference in point of view lies in the measure of our knowledge about the real world rather than an inherent property of the system considered. It was described as objective or subjective respectively in Chapter 10 and described by Haken in his synergistic approach [HAK 77]. For the statistical development of information in his Chapter 3, the author describes entropy as a "curse of statistical mechanics". From this analysis, it shows the difficulty of reconciling the two possible approaches, mechanical and thermodynamical. The generalization is known as statistical inference when traditional approaches are no longer effective. It is a set of approaches which involves the inducement of unknown characteristics of a population from a selected sampling by using deep learning, for example, in neural networks [BEN 16]. However, this mass processing of data poses questions about the reproducibility of the results.

When information is given or extracted from a system, there is communication with the external environment, even if it is only by thought as in the initial case of Maxwell's demon. As we saw in Chapter 5 (section 5.3.1), the demon can select the passage of molecules between two containers, which leads to a cognitive function. Such a characteristic can be understood and extended to economic and living systems, as Monod (Nobel Prize in Physiology, 1965) advocated, citing the role of allosteric enzymes [MON 70]. They are assumed to play the same role by introducing an internal cognitive value, i.e. genetic information in a cell factory [DAN 11]. Thus, the acquisition or erasure of information is likened to a change in microscopic states in a system that can be modified by correcting transcription errors and becoming intelligent by progressive learning. This is the beginning of sensory robots possessing artificial intelligence and the advent of computer languages for communication, leading to the development of cognitive sciences.

11.1.2.3. *Generalized concept of entropy and chaos*

Without detailing the various mathematical definitions of microscopic entropy and in particular its extension in economics [VEL 11], we must mention the generalization of the Boltzmann–Gibbs formula proposed by Tsallis [TSA 11]. In a Lévy statistical distribution based on maximum entropy production, this formula (equation [11.1]) is no more than a limiting case of a more general formulation. It leads to non-extensive physical variables such as statistical entropy in particular. This characteristic, applicable to other thermodynamic functions, no longer falls within the phenomenological framework that allowed us to initially define the different forms of energy (see Table 2.1). The Gaussian distribution or usual normal law no longer corresponds to the type of statistic involved and the limits of the approach based on a standard thermodynamic system, which contains short-range interaction particles, are reached. As a result, the difficulty comes from the analysis technique selected and the choice of the appropriate distribution function.

In a becoming-chaotic situation, the presence of a strange attractor arising from one of the possible scripts (see Chapter 2, section 2.1.4) is linked to Tsallis' expression of entropy production that allows the different possible routes to be classified [ROB 05]. Thus, initially, there is a period-doubling bifurcation occurring in biological systems, whereas quasi-periodicity is present for cardiac arrhythmias. As for intermittence processes, linked to large fluctuations in some physical variable, they would be present in dynamic networks in computer science.

11.2. Classification of materials and devices

The thermodynamic classification of materials is based on their presentation with different classes, as introduced in Figure 1.10. We will not discuss traditional materials, which are essentially structural materials, but the other main families with an increasingly elaborate specific functionality. There is an obvious technological contribution involving one or more energy transformations, which we will summarize without discussing the manufacture conditions or the life cycle of these objects.

11.2.1. *Functional advanced materials*

These materials combine a structural characteristic with a specific response. They are essentially materials that respond to a stress, typically of an electrical or optical nature, as described in Chapters 3 and 4. Note that they can also be porous and selectively absorb vapors or liquids according to specific interactions and imposed geometric criteria: size, shape and connectivity of pores. In this context, based on the thermodynamic response of the system, two points must be highlighted:

– the particular role of interfaces which, according to their specificity, determine the function of the material or device constituting a polyphasic assembly. The role of selectivity, with respect to the value of one of the intensive variables of an interface, is then paramount in the system hierarchy (see section 11.3.2);

– the system response can be linear in a situation close to equilibrium but a more sophisticated formulation further away from this equilibrium. For a linear response obeying Onsager's relations, it is associated with the great phenomenological transport laws. In particular, cross-effects in the presence of two potentials/voltages explain many physical properties (see Chapter 3). However, the discoveries of the transistor and the laser are significant examples of a nonlinear response which can be found beyond bifurcation points and even reach a chaotic state.

In Chapter 4, the conversion and storage of electrical energy, but also of information, was discussed. From different primary energies, possible conversions and storage of electricity are reviewed by examining the associated energy yields, the key to a technically viable solution. In addition, the physical storage of information is described from the notions of a hysteresis loop near equilibrium and a memory effect in the case of spatial bistability, derived from the thermodynamics of systems far from equilibrium.

11.2.2. *Nanomaterials and quantum mechanics*

The functioning of classical systems is based on wave particle identification and the nature of these particles (fermions or bosons) in quantum mechanics. However, the miniaturization that gave rise to nanomaterials allows us to reveal a previously underlying strictly quantum behavior (see Chapter 5). An experimental approach, in particular on new materials, results in active nanosystems in electronics, optics and magnetism, which are presented in Chapter 6. The ultimate concept of a single molecule is then considered, despite avoiding a strictly statistical description of the medium, which is replaced by a time average for a given stationary state (the ergodic hypothesis).

The most innovative point is that of quantum computers with the development of quantum memories on adapted physical supports. They must present a logic adapted to offer new perspectives currently in rapid evolution. A century after the advent of quantum physics, based on the wave-particle equation and the phenomenon of entanglement, the first applications of teleportation, coherent transport of a quantum state identified in space, were realized. This second quantum revolution will define the calculation rules and the possible structure of such a computer [DEU 16].

11.2.3. *Biomaterials inspired by living environments*

In Chapter 7, we developed three approaches: the mimicry of a living organism, the development of a hybrid system with the introduction of a living–inert interface and bioinformatics. The first approach relies on morphogenesis and shape analysis by comparing living and inert environments. The second approach gives rise to numerous sensors and other devices such as bioconverters that involve the notions of biocompatibility and active interface. The concept of energy efficiency and optimization of energy conversions becomes paramount as with conventional devices and machines. Finally, bioinformatics, which is based on the extension of information science in biology or bionics, appears in two aspects. On the one hand, it is guided by the use of enormous calculation capacities to simulate the functioning of living organisms and, on the other hand, synthetic biology operates by controlling or modifying the genetic code and the stored information which is ultimately transmitted. Thus, the semantic barrier between the inert and the living becomes more and more blurred.

To schematize this thermodynamic classification, we have summarized in Table 11.1 the main functionalities independent of the type of material used. They are all based on the operation of closed or open systems close to equilibrium.

Functionality thermodynamic classification	Sensors and detectors	Converters and transducers	Energy storage	Information storage
– Closed system in linear regime	Electronic probe Biosensors	Photovoltaic cells Light diodes, actuators MEMs, NEMs	Thermal (heat) Mechanic (inertia) Electrical (batteries)	Physical memories: electronic, optical, magnetic, biological (DNA)
– Open system in linear regime	Chemical sensors, selective adsorbers	–	Electrochemical (fuel cell), chemical reactors	–

Table 11.1. *Summary table of functional materials in a linear regime close to thermodynamic equilibrium*

11.2.4. *Extension to living organisms, ecological and economic systems*

The behavior of living environments beyond the molecular level, presented in Chapter 9, shows that these are systems that commonly work in dissipative regimes, i.e. in the vicinity of a bifurcation point. This is different to materials, which often work in a linear regime as we have just discussed. In the case of populations of individuals and ecological systems, the boundary between the system and its environment becomes blurred. For economic and financial systems, its rigorous definition is more debatable, as we have indicated with the role of currency as an indirect variable with a quasi-metaphorical character. However, their dynamic behavior satisfies the same sequences indicating a real analogy (see Figure 11.1). The main difference for living systems is that the system response is no longer constant for constraints and variable imposed flows. There is, therefore, a specificity in living beings presenting an intrinsic capacity to evolve as in the phenomenon of aging or during reproduction. Living species have their own internal information and are characterized by biological rhythms and characteristic cycles that we will analyze.

11.3. Rhythms, complexity and synergy of dynamic systems

The aim is to analyze and compare various thermodynamic systems as complex self-organized systems as defined by Haken [HAK 77]. This author introduced the notion of synergy when several factors act simultaneously to create a specific overall effect. We will discuss the developed aspects of materials, then extend it to living systems, and furthermore ecological and economic systems. The following points will be addressed:

– the analysis of shapes and the logical approach;

– dimensional scales and their hierarchical responses;

– the constraints and flows imposed on a system with characteristic periods and cycles;

– the global behavior of dynamic systems;

– a more general definition of complexity.

11.3.1. *From the analysis of shape to functionality*

As discussed in Chapter 1, the two approaches concerning the content and form are always present from the idealized points of view of Plato and Aristotle (see Figure 1.1). Geometric shapes, particularly regular polyhedra, have served as a basis for structural chemistry. In practice, the shapes of materials are derived from crystallogenesis and more specifically from morphogenesis, a concept extended to biology. Thus, Changeux discussed the analogy between "The Life of Forms and the Forms of Life" [CHA 12] by comparing functional forms that include anatomy, the artifact of living beings and purely aesthetic forms. In this analysis of shapes, the fractal geometry of Mandelbrot [MAN 74], which describes self-similar irregular forms generated in nature or on materials, is a significant recent advance. This recognition of shapes and their spatial arrangement is improved and extended with the development of computerized means to analyze them (see the growth of dendritic forms in Figure 2.13). The next step is the generalization of shapes considered as a dynamic organization of simple elements in space and time. They are classed as structured ensembles and not as a juxtaposition of elements: for example, a succession of static images does not constitute a film. Thus, experimental ensembles, developed as an oscillating electric circuit and formed of technical devices assembled using specific tools, have their own dynamics. For example, mathematical methods are used to control a smart grid that can be formed of coupled chaotic oscillators [LIN 16].

Meanwhile, a logical approach introduced by Aristotle has always been pursued. A significant event was "The Discourse on the Method" by Descartes published in 1637, "thinking is reasoning", followed by many other contributions in the philosophy of science. The conceptual development in this field of dynamic systems can be found in the 1970s. As we mentioned in Chapter 1 (see Figure 1.8), the books of Thom and Glansdorff-Prigogine were published at the same time. There is a more mechanistic and topological approach in parallel with Thom's catastrophe theory and Mandelbrot's fractal geometry [MAN 74]. However, a view based on thermodynamics' far from equilibrium states and the self-organization of the system after one or more bifurcation thresholds has a more general scope. It is complemented by the analysis of chaos with Ruelle's contribution on the concept of chance and determinism established at the same time [RUE 92]. This author introduced the concept of the strange attractor that Mandelbrot has likened to a fractal figure. All these approaches are based on the questioning of the reductionist hypothesis, which no longer explains the emergence of new properties, as described in a holistic approach defined when the whole is more than the sum of the parts. An example is given by Anderson in 1972 [AND 72] who attempted to establish a hierarchical structure of scientific disciplines when an entity of a discipline X requires the application of the laws established for Y. For example, physiology depends on biology, itself derived from chemistry which in turn depends on the physics of the condensed matter and the quantum particles present. In this hierarchy, the emergence of new functionalities associated with specific interactions is the guiding principle. The consequences of these developments linking scientific disciplines to one another and explaining the emergence of new structures and organizations are still a topical issue for the understanding of dynamic systems.

11.3.2. *Scale analysis and organizational hierarchy*

The dimension or scale of different systems is a key parameter. For materials, the successive levels of structural organization, then textural and morphological, were presented in the third part of Chapter 2. As we have seen, the first descriptive level, atomic or molecular, is essential in nanomaterials and in molecular electronics where quantum phenomena are omnipresent (see Chapter 5). Then, in conventional materials, there are two further descriptive levels, mesoscopic and then macroscopic, which are linked to their manufacturing processes. We saw in Chapter 6 that colloidal processes and spatial self-organization create an additional hierarchy. This particular microarchitecture of space is analogous to that existing in living environments. One example is the case of proteins that have several structural levels, from primary to quaternary, which condition their metabolism and biological properties such as allosteria [MON 70].

From a functional point of view they are formed of subsystems which can be defined by their specific contribution to a device or a machine. Such a multiphase assembly will give rise to nonlinear behavior as with diodes and transistors. By creating a hierarchical structure in an electrical or optical circuit consisting of these elements and forming a sustained oscillator, a chaotic response can be observed.

In the case of living systems, a similar description can be undertaken as we have just recalled at the molecular level with proteins. At the top level, a cell (see Figure 9.1), which we consider as a homogeneous medium, is actually composed of a set of specific entities. Cell differentiation and organ formation are an additional step. At these different levels of description, it is necessary to associate nonlinear responses with a feedback loop and a behavior described in terms of dissipative regimes presenting a spatial or temporal self-organization.

Finally, by analogy, we have attempted to present macroeconomic systems as thermodynamic assemblies that may show a nonlinear response with a bifurcation threshold that is difficult to characterize. This approach experiences the problem of separating into financial and other interactive subsystems in order to establish an internal hierarchy.

11.3.3. *Constraints and flows: characteristic oscillations and cycles*

Recall that constraints are boundary conditions imposed on the system or on one of its components in the case of subsystems. As we have already introduced (see Figure 3.1), they are defined by the local values of the associated intensive thermodynamic variables. The quantitative aspect of exchanges or flows will condition the system response, linear or otherwise, and the notion of instability described by a bifurcation threshold. They give rise to temporal structures or to a self-organization of the system. The key point to clarify is that of the ability to detect or estimate, for a given critical flow, when the first system instability point appears. It is reflected, in particular, by the appearance of oscillations and characteristic cycles, for example in reaction–diffusion systems. The oscillatory period or characteristic frequency is a signature of behavior as in the chemical reaction of Belousov–Zhabotinsky (see Figure 2.6). By generalizing these cyclic oscillations or rhythms of living environments, we see that they are based on the competition between two antagonistic effects. Comparative examples are given in Table 11.2 below for different disciplines, demonstrating the generality of this phenomenon. For living species and geological, ecological or even economic systems, we have seen that different cycles coexist and have to match.

Time scales can vary enormously as well as dimensional scales and their possible synchronizations are characteristic of complex systems. The fact that it is difficult to exploit these oscillatory behaviors in thermodynamic terms results in the definition of existence diagrams in the system for the state considered and the quantification of the different flows. In particular, as we have reported, the transition from an oscillating state to a spatially bistable state with a memory effect is tricky to control.

Disciplines	Competitive phenomena	Oscillating systems and cycles
Physics	Coupling of oscillators and friction coefficients	Electric and parametric oscillator (Chua circuit)
Chemistry	Chemical reactions–molecular diffusion	Belousov–Zhabotinsky oscillating reaction
Biology	Autocatalytic genes activator–inhibitor	Circadian rhythms and biological clocks
Ecology	Predator–prey competition	Lotka–Volterra model
Economics	Supply–demand law	Short Kitchin cycle

Table 11.2. *Comparison of different far from equilibrium systems, based on the competition between two opposite effects causing the appearance of spatiotemporal phenomena*

11.3.4. *Dynamic and cybernetic systems*

A dynamic system is considered as a set of elements in time-dependent interactions. It has a causal and deterministic nature and can be modeled by a set of differential equations. If we know the exact initial state of a system, it is theoretically possible to exactly predict its temporal evolution (see Laplace's classical theory). As we introduced in Chapter 2, our limited initial knowledge leads to unpredictability and chance [RUE 92]. This limitation forms the basis of unstable dynamics, as Poincaré showed in his analysis of celestial mechanics. The emergence of deterministic chaos is the consequence; it is associated with the presence of a stable attractor basin and the characterization of an attractor [EKE 95]. As we have pointed out in the first part of Chapter 5 (note on Kolmogorov–Arnold–Moser theorem) only a fraction of the space phase is attainable for the evaluation of entropy production. A historical example of this type of attractor is that present in the solar system during its chaotic formation [LAS 89], which is at the origin of certain climatic phenomena (see Chapter 9). Thus, as we have already indicated, a turbulent regime in a weather forecast can lead to a chaotic situation. It is a Lorenz attractor, of which a representation of the typical trajectories in the phase space forms a double spiral with the shape of butterfly wings, as shown in Figure 11.2.

Figure 11.2. *Numerical simulation of trajectories confined in the three-dimensional phase space for the strange attractor of Lorenz's meteorological model, which gave rise to the butterfly effect*

Moreover, external stimuli can disrupt the onset or stability of this chaotic state. This is the case for vital organs, such as the heart or the brain, but also for stock markets. We must consider the possibility of a critical stimulus, depending on its nature, its intensity and its synchronism, modifying the response of the system and producing the appearance of a new behavior. In particular, in the study of these dynamics, its piloting involves the control and communication of information that reflects the existence of a feedback loop that we have just discussed. This is known as cybernetics or piloting science, introduced by Wiener, focused on the regulation and control of artificial or living systems. This discipline, based on human–machine communication, has not only developed in the fields of automatism and computer science but also in artificial intelligence thanks to the cognitive approach [WEI 48]. At present, the theoretical regulation of complex systems by algorithms control their dynamics and allow a solution to be chosen according to criteria established in advance, thus leading to decision making. As we have already shown they must be tackled using a systemic approach [DER 95]. Their modeling is no longer merely analytical, but requires an operational hierarchy of activities leading to a decision analysis [LEM 99]. Models of complex networks (see Box 9.1), with interdependent oscillators that can be deployed in a chaotic situation, are developed to account for them. Controlling the synchronization of the elementary activities appears essential for the stable and robust operation of a complex system composed of coupled oscillators [LIN 16].

11.3.5. *Toward a definition of complex systems*

These are systems where a new approach allows the study of relationships between the different parts or subsystems. It gives rise to an emergent collective behavior in interaction with the environment. Usually, different kinds of complexities are distinguished, i.e. chemical, biological, ecological, economic or even sociological. We will focus mainly on the physical and chemical [WHI 99], then biological [RIC 99] complexities.

The transition from shape to functional complexity is linked to functional hierarchization and an increase in degrees of freedom in an inert system. Let us recall the main levels from electronic components:

– materials (semiconductors, metals) and associated devices (diodes, transistors, etc.);

– electronic circuits and physical memories (smart cards, microprocessors);

– computer architecture and mass data acquisition ("big data");

– intelligent systems and robots, neural networks and deep learning (artificial intelligence).

For living beings, this stage is much more common; it gives rise to emergent physiological characteristics which no longer result from the addition of components (see the concept of holism). There is then the transmission of structural and functional information of a reproductive or evolutionary nature. There is a link between the appearance of self-organization and information contained in a message: this is biological complexity [RIC 99]. It therefore appears that the amount of information contained in a cell is much greater than the strictly genetic information. For example, the cytoskeletal self-assembly role is predominant in cell functions with active metabolism far from equilibrium [HUB 13]. Biological systems contain more information in the genetic code than that required for the biosynthesis of the elements involved in a successive set of scaffolds. It is an autonomous complex system that can be summarized by a set of propositions that is similar to the different levels in electronic systems:

– it consists of individual elements or operators acting on the environment and connected to each other;

– such a system, closed or open, possesses a certain structural and functional organization or hierarchy and can present nonlinear responses with control loops and threshold effects;

– the system is in a dynamic state with its own history, its state at a given instant may depend on previous situations;

– its properties and characteristics are emerging; they are no longer predictable from the elements or subassemblies that compose it;

– several factors intervene to create synergy or symbiosis [HAK 77]; there are one or more dynamic and recursive causes (loop effect) and a simple causal direct explanation is no longer satisfactory (but there are synergistic correlations);

– the local thermodynamic equilibrium approach includes simultaneous flows of energy, matter or information that can be continuous or sporadic (stimuli) and modify the system response.

The purpose of living organisms is to adapt to the environment. They acquire usable information and have internal autonomy during the reproductive phase since they are causally closed [PER 13b]. This intrinsic information might be represented as we have seen by the presence of a Maxwell demon, thus giving it an ontological character [MON 70].

Note that in this complex metabolic process, the role of cycles and temporal rhythms appears to be fundamental. Their synchronization associated with a hierarchy of behaviors now appears crucial in a cyclical process of time. It is the myth of the great watchmaker, present in Voltaire among others, which is revisited.

11.4. Epilogue: descriptive uniqueness and limitation of thermodynamic bases

Some key points to remember are the uniqueness of the approach and its limitations:

– a material, whatever its origin, is defined as the combination of intrinsic volume properties and a surface or interface characteristic controlling the nature and quantity of the various exchanges; in nanomaterials, this interactive surface is favored. Considered as a system in interaction with the environment, its behavior is classified according to thermodynamic criteria generally situating it in the vicinity of equilibrium. Some more elaborate devices may, however, reach as far as a nonlinear response and a chaotic situation. An important consequence is the classification of transformation functions and the optimization of energy conversions to obtain and rationally use a vital secondary energy such as electricity. This is a key aspect in economics, especially when sacrificing the thermodynamic yield in favor of the power delivered. They are then governed by the temporal irreversibility of a physicochemical process and the principle of maximal entropy production;

– mimicry with in living environments often guides these transformations. The extension to living systems starting from molecular biology is effective and successively extends to living beings and their populations within the context of ecology. For example, natural and artificial cycles have the same spatiotemporal characteristics. Their frequency and synchronization then appears vital for these complex hierarchical systems. However, despite the observed analogies based on the notions of synergy and hierarchy [HAK 77], a completely thermodynamic approach still poses problems. In particular, the modeling of such systems with the definition of relevant extensive and intensive local variables is currently an obstacle in a system that must respond to different types of stimuli [LEM 99]. This is the case, for example, in a human brain subjected to electrical impulses, showing the synaptic plasticity of the neural network, or in a financial system such as a stock market, when many external disturbances are induced numerically by the set of fast algorithms. The system response time becomes relevant and the arrival of information can even generate "burn-out", which modifies its response and can induce another type of behavior leading to a chaotic situation;

– the limits of the classical thermodynamic model exist, especially for nanosystems, when considering few particles often in a purely quantum regime or in the presence of rare events that no longer belong to a Gaussian distribution. Moreover, the transition from a random stochastic probability to the taking into account of acquired information is often introduced as a Maxwell's demon. The latter induces a conditional probability, which is a significant step since the system then acquires memorized information and a cognitive character. The analysis then changes from an objective approach to a subjective or Bayesian inference [DEW 09b]. It is the transition from storage and reading on a physical medium to the analysis of information contained in the message by the observer. This decisive passage is translated by Morin, who examines acquired and transmitted knowledge as information that has a meaning in a given context or a cognitive environment. This element is essential for the elaboration of a complex transdisciplinary thought which then has a sociological aspect [MOR 90].

Glossary

These definitions concern the words and acronyms often used in the text with the meaning attributed to them. In doing so, we do not escape the vicious circle of defining the primitive words in relation to each other.

Adsorption: Fixation of atoms or molecules onto the surface of a solid by different mechanisms; we usually distinguish the processes of physi- and chemisorption.

Algorithm: A set of mathematical rules whose application solves a problem stated by means of a finite number of instructions in a computer programming language.

Artificial intelligence: Computer devices imitating or replacing the human for certain implementations of its cognitive functions (confer the Turing test).

Biomaterial: Inert material used in a biological or medical device and designed to interact and be tolerated by living media.

Bionic: The science of systems whose mechanisms of operation are copied from those of natural environments.

Biosphere: Parts of Earth in which life is permanently possible and which brings together all living beings.

Colloids: Suspension of very small particles, at the nanometric scale, in a fluid medium with a homogeneous stable appearance.

Complexity: A new feature that is no longer merely the sum of components present in the system under consideration.

Constraint: Permanent stress to which a system is subjected via the control of an intensive variable (temperature, pressure or chemical potential).

Converters: Transform primary energy (heat, work, radiation, chemical reaction) into electricity, easily usable secondary energy.

Cristallogenesis and morphogenesis: Process of formation and evolution of condensed matter. Crystallization is the ordered assembly of atoms or molecules forming a crystal lattice with translational symmetry at the microscopic level. Morphology accounts for textures and forms obtained on meso- and macroscopic scales; the extension to living beings is a so-called autonomous or directed morphogenesis.

Deterministic chaos: Non-periodic oscillating behavior of a dynamic system far from thermodynamic equilibrium; a concept linked to that of a strange attractor.

DNA: Deoxyribonucleic acid composed of four nucleotides (adenine-cytosine and guanine-thymine pairs) and attached by phosphodiester bonds; it is made up of two strands wound in a helix whose pairs of nucleotides form the steps.

Ecosystem: Living space defined on land (part of the biosphere) where the interactions between different living organisms and with their environment are analyzed.

Energy: Fundamental physical quantity defined as the product of an intensive thermodynamic variable by an extensive variable; it comes in different forms, the main ones are of thermal, mechanical, radiative, electrical, chemical and nuclear origin.

Entropy: A thermodynamic quantity, a non-conserved magnitude, which is a measure of the degradation of energy in the form of heat during the irreversible evolution of a system.

Equilibrium state, stationary state and transitory state: The state of equilibrium is a thermodynamic situation corresponding, respectively, to a stable state of minimal energy, which is independent of time; in the presence of a constraint or an external flux, the responses of the system have stationary or time-dependent values, it is a permanent or even oscillatory regime; in the case where the response time of the system conditions an external disturbance, it is a transient regime.

Exergy: Fraction of the usable energy during conversion or conversion; this quantity depending on the experimental conditions cannot be preserved.

Flux: Continuous transfers or exchanges of energy, information or matter that are permitted between a defined system and part of its environment acting as a reservoir.

Fractal: A rough, porous or fragmented object at all observed dimensions; it is a property of self-similarity meaning that their geometry is invariant by a change in scale.

Genes and genomes: A sequence of nucleotides inscribed in DNA giving the genetic code and containing the information necessary for the synthesis of RNA (ribonucleic acid) and protein molecules; the entire DNA of a living organism forms the genome.

Genetics and epigenetics: Study of the hereditary characteristics materialized by the genetic code and its transmission ensured by the replication of the DNA; epigenetics is the study of the mechanisms that modulate the expression of the genetic heritage according to its environment.

Hysteresis: Characteristic of a system close to equilibrium presenting a property at a given instant depending on its previous evolution; it is also the response to an external constraint far from equilibrium with a certain delay and can then be responsible for a nonlinearity and a bistability. A physical hysteresis may be of mechanical, thermal, electrical, optical or magnetic origin.

Information: It is a set of data obtained through perception, an observation or signal from a sensor and have a meaning for an observer in a particular field.

Information theory: Initiated by Shannon, it is a probabilistic mathematical model that makes it possible to quantify the content of information; it is connected to thermodynamics (statistical concept of entropy) and in principle requires usually a minimum of energy to create, store, read and erase a message.

Lithography: Printing process to create and to copy an image, initially drawn on a stone; in electronics generalized to any chemical process allowing to realize a printed microchip.

Material: Solid phase, possibly an organized liquid, with a geometrical shape and special properties with a surface functionality enabling it to specifically exchange with its environment.

Metabolism: All of the coupled biochemical reactions occurring in the cells of a living organism and conditioning its behavior.

Metamaterial: Heterogeneous material with a periodic structure and presenting peculiar electromagnetic properties, for example, a negative refractive index.

Model: Schematic representation of a physical phenomenon by reducing the number of significant variables and establishing mathematical relations between them.

Nanomaterial: A material with at least one nanometric dimension and thereby having an increased surface or interface for a given volume.

Networks: Set of discrete entities represented by nodes and links that connect them together according to a defined topology or architecture (see percolation models and neural networks).

Neuron: Naturally, it is an excitable living cell ensuring the transmission of an electrical signal called nerve impulses; artificially, this is an electronic device simulating its biological functionality.

Object: Entity defined in a three-dimensional space made by man; perceptible form through sight or touch that can be aesthetic or functional.

Observation scales and dimensions: In a material, the scales of sizes or dimensions are usually called micro-, meso- and macroscopic and the associated structural, textural and morphological characteristics; microscopic involves those up to a nanometer, mesoscopic include those between the nanometer and the micron, and finally macroscopic is anything beyond the micron until direct visual observation.

Particle: A small definite portion of matter, which has become a fundamental physical quantity in quantum mechanics and high-energy physics.

Percolation: Physical phenomenon with a critical behavior that is represented by a mathematical theory of a discrete lattice to describe the characteristics of random heterogeneous media for modeling transport or propagation.

Phase space: Abstract space represented by the dynamic variables, positions and quantities of motion of each particle forming a given system.

Processor and microprocessor: A central information processing unit that executes a logical series of instructions stored in a computer program.

Proteins: Biological macromolecules consisting of amino acid residues; their sequence and length in the peptide chain as well as their multiscale structure at different levels defined their specific properties; in particular, they are essential for interpreting the genetic code and in their role as enzymatic catalyst.

Sensors: Device for detecting a specific signal usually of an electrical or optical nature.

Stimulus: External action acting punctually with a sufficient intensity (sensitivity threshold) at a given moment on a system and causing a physical, chemical, biological or even physiological response.

Stochastic: Characterizes a time-dependent random phenomenon that lends itself to statistical analysis out of equilibrium conditions at the microscopic level.

Strange attractor: A region of the phases hyperspace called attraction basin where a dynamic system can enter but can no longer leave it.

Substance and matter: We use the term substance in the primitive sense of an unspecified entity that exists in itself with a permanent reality and the word matter in a more concrete sense presenting a tangible character, going toward the notion usually used in physics (states of matter) and the derived notion of material.

Synergy: Several factors act together to create a particular overall effect; this is the case in an interdisciplinary approach involving several causes.

Thermodynamic phase: Homogeneous assembly of atoms or identical molecules occupying a meso- or macroscopic volume in space and defined in an equilibrium or existence diagram.

Thermodynamic system (isolated, closed, or open): In thermodynamics, a system defined by an enclosure or wall is said to be isolated if it has no exchange with the external environment; a closed system only exchanges energy and an open system of energy and matter: only the last two corresponding to non-equilibrium states of the system are of interest in this study.

Topology: Study of the spatial deformations of an object by continuous transformations; it allows a mathematical classification in topological spaces.

Transducer: An inverting device of a converter, transforming an electrical signal into another form of energy; an actuator is a special case transforming electrical source into mechanical energy.

Trophicity: The mechanisms and processes involved in the nutrition and growth of organs and tissues.

Bibliography

[ADA 90] ADAMSON A.W., *Physical Chemistry of Surfaces*, 4th ed., John Wiley and Sons, New York, 1990.

[ADA 02] ADAM J.-L., "Verres et polymères pour l'optique guidée", *L'actualité chimique*, vol. 3, pp. 111–115, 2002.

[ADD 13] ADDO P.M., BILLIO M., GUEGAN D., Nonlinear dynamics and recurrence plots for detecting financial crisis, Documents de travail du centre d'économie de la Sorbonne, Paris, available at : https://halshs.archives-ouvertes.fr/halshs-00803450, 2013.

[ALA 14] ALATRISTE CONTRERAS M.G. and FAGIOLO G., "Propagation of economic shocks in input-output networks: a cross-country analysis", *Physical Review E*, vol. 90, 062812, 2014.

[ALL 01] ALLEAU T., DUMAS I., "La pile à combustible: une réponse aux préoccupations énergétiques et environnementales", *Clefs CEA*, vol. 44, pp. 42–46, 2000–2001.

[AMA 11] AMATORE C., ARBAULT S., GUILLE M., LEMAITRE F., "Synapses artificielles et stress oxydant", *L'actualité chimique*, vol. 348–349, p. 2531, 2011.

[ANC 02] ANCEAU F., *Histoire des machines informatiques, Qu'est-ce que les technologies*?, Odile Jacob, Paris, 2002.

[AND 72] ANDERSON P.W., "More is different", *Science*, vol. 177, no. 4047, pp. 393–396, 1972.

[AND 15] ANDRÉJAK C., DELHAES L., "Le microbiome pulmonaire en 2015: une fenêtre ouverte sur les pathologies pulmonaires chroniques", *Médecine/Sciences*, vol. 11, pp. 971–978, 2015.

[ARG 88] ARGOULT F., ARNEODO A., GRASSEAU G. *et al.*, "Self-similarity of diffusion limited aggregates an electrodeposition clusters", *Physical Review Letters*, vol. 61, no. 22, pp. 2558–2561, 1988.

[ARN 98] ARNÉODO A., MUZY J.-F., SORNETTE D., "Direct causal cascade in the stock market", *The European Physical Journal B*, vol. 2(2), pp. 277–282, 1998.

[ART 11] ARTERO V., GUILLET N., FRUCHART D. *et al.*, "L'hydrogène, une énergie propre pour demain?", *Pour la Science*, vol. 405, pp. 26–34, 2011.

[AUZ 14] AUZANNEAU M., "Gaël Giraud : 'Le vrai rôle de l'énergie va obliger les économistes à changer de dôgme'", available at : petrole.blog.lemonde.fr/2014/04/19, 2014.

[AVI 74] AVIRAM A., RATNER M.A., "Molecular rectifiers", *Chemical Physics Letters*, vol. 29, pp. 277–283, 1974.

[AYR 84] AYRES R.H., NAIR I., "Thermodynamics and economics", *Physics Today*, vol. 37, pp. 62–71, 1984.

[BAL 14] BALL P., "La mystérieuse boîte noire de D-wave", *La Recherche*, vol. 485, pp. 26–33, 2014.

[BAR 01] BARDOU F., BOUCHAUD J.-P., ASPECT A. *et al.*, *Lévy Statistics and Laser Cooling: How Rare Events Bring Atoms to Rest*, Cambridge University Press, Cambridge, 2001.

[BAR 13] BARNOUS L., "L'hypothèse de la soupe primitive revisitée", *Les dossiers de la Recherche*, vol. 2, pp. 44–47, 2013.

[BAR 15] BARGE L.M., SILVANA S.S.C., CARTWRIGHT J.H.E., "From chemical gardens to chemiobrionics" *Chemical Review*, vol. 115, pp. 8652–8704, 2015.

[BAR 16] BARAN D., ASHRAF R.S., HANIFI D.A., "Reducing the efficiency-stability-cost gap of organic photovoltaics with highly efficient and stable small molecule acceptor ternary solar cells", *Nature Materials*, 2016.

[BAS 10] BASS J., TAKAHASHI J.S., "Circadian integration of metabolism and energetics", *Science*, vol. 330, pp. 1349–1354, 2010.

[BAU 04] BAUDET J., *Penser la matière*, Vuibert editor, Paris, 2004.

[BEC 02] BECK G., "Les matériaux: du fondamental aux applications", *L'actualité Chimique*, special issue. 3, pp. 10–12, 2002.

[BEG 06] BEGUIN F., YAZAMI R., "Les carbones dans les systèmes de stockage électrochimique de l'énergie", *L'actualité chimique*, vol. 295–296, pp. 86–90, 2006.

[BEN 00] BEN-JACOB E., COHEN J., LEVINS H., "Cooperative self-organization of micro-organisms", *Advances in Physics,* vol. 49, no. 4, pp. 395–554, 2000.

[BEN 03] BENNETT C.H., "Notes on Landauer's principle, reversible computation and Maxwell's demon", *Studies in History and Philosophy of Sciences*, vol. 34, pp. 501–510, 2003.

[BEN 13] BENOIT P., VEDRAL V., "L'ordinateur quantique", *Le cahier du Monde des Sciences*, vol. 1, pp. 65–72, 2013.

[BEN 16] BENGIO Y., "La révolution de l'apprentissage profond", *Pour la Science*, vol. 465, pp. 42–49, 2016.

[BER 84] BERGE P., POMEAU Y., VIDAL C., *L'ordre dans le chaos: vers une approche déterministe de la turbulence*, Hermann, Paris, 1984.

[BER 85] BERTHELOT M., *Les origines de l'Alchimie*, Georges Steinheil, Paris, 1885.

[BER 07] BERGSON H., *L'évolution créatrice*, 1st edition, PUF, Paris, 1907.

[BER 09] BERRY H., CESSAC B., "Du chaos dans les neurones", *Pour la Science,* vol. 385, pp. 309–305, 2009.

[BER 12] BERUT A., ARAKELYAN A., PETROSYAN A. *et al.*, "Experimental verification of Landauer's principle linking information and thermodynamics", *Nature*, vol. 483, pp. 187–189, 2012.

[BER 13] BERGGREN G., ADAMSKA A., LAMBERTZ C. *et al.*, "Biomimetic assembly and activation of [Fe-Fe]-hydrogenases", *Nature*, vol. 499, pp. 66–69, 2013.

[BOC 68] BOCCARA N., *Les principes de la thermodynamique classique*, PUF, Paris, 1968.

[BOI 09a] BOISSONADE J., DE KEPPER P., "Chemomechanics: oscillatory dynamics in chemoresponsive gels", in BORCKMANS P., DE KEPPER P., KHOLHOV A.R. *et al.* (eds), *NATO-ASI Series*, pp. 95–116, Springer Science, Dordrecht, The Nederlands, 2009.

[BOI 09b] BOISSONADE J., DULOS E., DE KEPPER P., "Des structures de Turing à la chimiomécanique", *L'actualité chimique*, vol. 336, pp. 17–21, 2009.

[BON 01] BONN D., "Wetting transitions", *Current opinion in Colloids and Interface Science*, vol. 6, pp. 22–27, 2001.

[BON 05] BONTEMPS N., SIMON C., "Supraconducteurs à haute température critique vingt ans après", *Images de la Physique*, CNRS, Paris, pp. 98–104, 2005.

[BON 13] BONNET J., YIN P., ORTIZ M.E. *et al.*, "Amplifying genetic logic gates", *Science,* vol. 342, no.6132, pp. 599–603, 2013.

[BOR 05] BOREL L., FAURAT D., *Thermodynamique et énergie, de l'énergie à l'exergie*, Presses polytechniques et universitaires Romande, Lausanne, 2005.

[BOU 04] BOUCHAUD J.-P., POTTERS M., *Theory of Financial Risk and Derivative Pricing*, 2nd edition, Cambridge University Press, 2004.

[BOU 80] BOULIGAND Y., *La morphogénèse de la biologie aux mathématiques*, Maloine, Paris, 1980.

[BOU 84] BOURDON J., "Matériaux pour l'enregistrement de l'information", *L'Actualité chimique*, vol. 12, pp. 15–29, 1984.

[BRE 13] BREAKSPEAR M., "Generative models of cortical oscillations: neurobiological implications of the Kuramoto model", *Human Neuroscience*, available at: http://dx.doi.org/10.3389/fnhum.2010.00190, 2013.

[BRI 59] BRILLOUIN L., *La science et la théorie de l'information*, Masson, Paris, 1959.

[BRO 07] BROOKS K.P., HU J., ZHU H., "Methanation of carbon dioxide by hydrogen reduction using Sabatier process in microchannel reactors", *Chemical Engineering Science,* vol. 62, no. 4, pp. 1161–1170, 2007.

[BUS 05] BUSSEREAU P., ROUX A., "Formation et dynamique de nanotubes de membranes", *Images de la physique,* vol. 2005, pp. 245–250, 2005.

[BYE 02] BYENUS J.M., *Biomimicry: Innovation Inspired by Nature*, William Horrow Paperbacks, New York, 2002.

[CAI 06] CAIN M.L., DAMMAN H., LUE R.A. *et al.*, *Découvrir la Biologie*, De Boeck, Brussels, 2006.

[CAL 85] CALLEN H.B., *Thermodynamics and an Introduction to Thermostatistics*, 2nd ed., John Wiley and Sons, New York, 1985.

[CAL 14] CALOVI D.S., LOPEZ U., NGO S. *et al.*, "Swarming, schooling, milling: phase diagram of a data-driven fish school model", *New journal of Physics*, vol. 16, p. 015026, 2014.

[CAR 82] CARTER F.L., *Molecular Electronic Devices*, 1st ed., M Dekker, Amsterdam, 1982.

[CAS 70] CASTAING R., *Cours de thermodynamique statistique*, Masson, Paris, 1970.

[CHÂ 05] CHATELAIN F., FUCHS A., "Les micro et nanotechnologies pour le vivant", *Clefs CEA*, vol. 52, pp. 102–106, 2005.

[CHA 12] CHANGEUX J.P., *La vie des formes et les formes de la vie*, Odile Jacob, Paris, 2012.

[CHE 14] CHENG Q., JIANG L., TANG Z., "Bioinspired layered materials with superior mechanical performance", *Accounts of Chemical Research*, vol. 47, no. 4, pp. 1256–1266, 2014.

[CHE 17] CHENG C., LIU S., SHI X.-Q. *et al.*,"Weak synchronization and large-scale collective oscillation in dense bacteria suspensions", *Nature*, vol. 542, pp. 210–214, 2017.

[CHO 16] CHONG M.C., SOSA-VARGAS L., BULOU H. *et al.*, "Ordinary and hot electroluminescence from single-molecule devices: controlling the emission color chemical Engineering", *Nano Letters*, vol. 16, no. 10, pp. 6480–6484, 2016.

[CHU 94] CHUA L.O., "Chua's circuit: an overview ten years later", *Journal of Circuits, Systems and Computers*, vol. 4, pp. 117–159, 1994.

[CLA 10] CLARK J., LANZANI G., "Organic photonics for communications", *Nature Photonics*, vol. 4, pp. 438–446, 2010.

[CLE 17] CLERC D., "Cycles économiques", *Encyclopedia Universalis*, available online, 2017.

[COI 03] COIRAULT C., POURNY J.-C., LAMBERT F. *et al.*, "Les pinces optiques en biologie et médecine", *Médecine et Sciences*, vol. 19, pp. 364–367, 2003.

[COL 08] COLFEN H., ANTONIETTI M., *Mesocrystals and Non-Classical Crystallization*, John Wiley and Sons, New York, 2008.

[CON 09] CONT R., "La statistique face aux évènements rares", *Pour la Science*, vol. 385, pp. 116–123, 2009.

[COV 14] COVERT M., "Simulating a living cell", *Scientific American*, vol. 310; no. 1, pp. 44–51, 2014.

[DAN 10] DANINO T., MONDRAGON-PALOMINO O., TSIMRING L. *et al.*, "A synchronized quorum of genetic clock", *Nature*, vol. 463, pp. 326–330, 2010.

[DAN 11] DANCHIN A., "Les gènes du démon de Maxwell: est-il possible de construire une usine cellulaire?", *Comptes rendus chimie de l'Académie des sciences*, vol. 14, no. 4, pp. 413–419, Paris, 2011.

[DAU 13] DAUBIN V., PENEL S., TANNIER E., "L'ADN mémoire numérique du vivant", *Pour la Science*, vol. 433, pp. 102–108, 2013.

[DEB 16] DE BERARDINIS R.J., CHANDEL N.S., "Fundamentals of cancer metabolism", *Science Advances*, vol. 2, no. 5, p. e1600200, 2016.

[DEG 76] DE GENNES P.G., "Transition de percolation", *La Recherche*, vol. 72, pp. 919–927, 1976.

[DEH 14] DEHAENE S., “Science avec conscience”, *La Recherche*, vol. 492, pp. 66–69, 2014.

[DEK 09] DE KEPPER P., BOISSONADE J., SZALAI I., *Chemochemical Instabilities in Responsive Materials*, Springer, Dordrecht, 2009.

[DEL 03] DELHAES P., “Attempts to chemical vapour infiltrate pyrocarbons: evidence of a spatial bistability?” *Carbon*, vol. 41, pp. 1093–1095, 2003.

[DEL 06] DELHAES P., OLRY P., “Fibres de carbone et matériaux composites”, *L'actualité chimique*, vol. 295–296, pp. 42–46, 2006.

[DEL 11a] DELHAES P., *Carbon Based Solids and Materials*, ISTE Ltd, London and John Wiley and Sons, New York, 2011.

[DEL 11b] DEL RIO L., ABERG J., RENNER R. *et al.*, “The thermodynamic meaning of negative entropy”, *Nature*, vol. 474, pp. 61–63, 2011.

[DEL 12] DELHAES P., “Carbon science and technology: from energy to materials”, ISTE Ltd, London and John Wiley and Sons, New York, 2012.

[DEL 17] DELAHAYE J.-P., “Le vivant plus fort que l'électronique”, *Pour la Science*, vol. 473, pp. 78–83, 2017.

[DEL 90] DELHAES P., “Review on the radical ion salts”, in METZGER R.M., DAY P., PAPAVASSILIOU G.C. (eds), *Lower Dimensional Systems and Molecular Electronics*, Plenum Press, New York, 1990.

[DER 95] DE ROSNAY J., *L'homme symbiotique*, Le Seuil, Paris, 1995.

[DES 09] DESAI R.C., KAPRAL R., *Dynamics of Self-organized and Self-assembled structures*, Cambridge University Press, Cambridge, 2009.

[DES 13] DE S.CAVALCANTE H.L.D., ORIA M. *et al.*, “Predictability and suppression of extreme events in a chaotic system”, *Physical Review Letters*, vol. 111, no. 19, p. 198701, 2013.

[DEU 16] DEUTSCH D., EKERT A., “S'affranchir des limites du quantique”, *Dossier pour la science*, vol. 93, pp. 26–30, 2016.

[DEW 05] DEWAR R.C., “Maximum entropy production and non-equilibrium statistical machanics”, in KLEIDON A., LORENTZ R.D. (eds), *Non-Equilibrium Thermodynamics and the Production of Entropy*, Springer-Verlag, Berlin, 2005.

[DEW 09a] DE WITT A., NICOLIS G., “La chimie non linéaire: acquis et promesses”, *L'actualité chimique*, vol. 336, p. 1016, 2009.

[DEW 09b] DEWAR R.C., “Maximum entropy production as an inference algorithm that translates physical assumptions into macroscopic predictions: don't shoot the messenger”, *Entropy*, vol. 11, pp. 931–944, 2009.

[DIN 05] DING H.F., SCHMID A.K., Li D. *et al.*, "Magnetic bistability of Co dots", *Physical Review Letters*, vol. 94, p. 157202, 2005.

[DIN 16] DING X., HE Y., DUAN Z.-C. *et al.*, "On-demand single photons with high extraction efficiency and near unity indistinguishability from a resonantly driven quantum dot in a micropiullar", *Physical Review Letters*, vol. 116, p. 020401, 2016.

[DOM 16] DOMINGUES DOS SANTOS F., "Les polymères fluorés électroactifs pour l'électronique imprimée", *L'actualité chimique*, vol. 403, pp. 63–64, 2016.

[DRE 12] DRECHSLER R., WILLE R., CHATTOPADHYAY S. *et al.*, "Reversible circuits: recent accomplishments and future challenges for an emerging technology", in RAHAMAN H. *et al.* (eds.), *International Symposium on VLSI Design and Test*, Springer, Berlin, 2012.

[DRO 15] DROZDOV A.P., EREMETS M.I., TROYAN I.A. *et al.*, "Conventional superconductivity at 203K at high pressures in the sulfur hydride system", *Nature*, vol. 525, pp. 73–76, 2015.

[DU 93] DU TRÉMOLET DE LACHEISSERIE E., *Magnetostriction: Theory and Applications of Magnetoelasticity*, CRC Press, New York, 1993.

[EAS 09] EASTES R.E., DARRIGAN C., BATAILLE X., "Les jardins chimiques: un faux pas vers la vie synthétique", *Pour la Science*, vol. 375, pp. 84–87, 2009.

[EIG 77] EIGEN M, SCHUSTER P., "The hypercycle, principle of natural self-organization", *Naturwissenschaften*, vol. 64, pp. 541–565, 1977.

[EKE 95] EKELAND I., *Le chaos*, Dominos Flammarion, Paris, 1995.

[ELG 13] ELGRISHI N., ARTERO V., FONTECAVE M., "Activation du dioxyde de carbone: enzymes, catalyseurs bioinspirés et photosynthèse artificielle", *L'actualité chimique*, vol. 155,pp. 95–100, 2013.

[ERM 15] ERMANN L., ERMANN K.M., FRAHM K.L. *et al.*, "Google matrix analysis of directed networks", *Review of Modern Physics*, vol. 87, pp. 1261, 2015.

[ETI 08] ETIENNE S., LAURENT D., GAUDRY E. *et al.*, *Les matériaux*, Dunod, Paris, 2008.

[EVA 01] EVANS A.G., "Lightweight materials and structures", *MRS Bulletin*, vol. 26, no. 10, pp. 790–797, 2001.

[FEN 13] FENG X., MATHONIERE C., JEONG L. *et al.*, "Tristability in a light-actuated single molecule magnet", *Journal of Chemical Society*, vol. 135, pp. 15880–15884, 2013.

[FER 01] FERT A., PETROFF F., "L'électronique de spin", *Images de la Physique*, vol. 2000, pp. 74–80, 2001.

[FER 07] FEREY G., "Les nouveaux solides poreux ou le miracle des trous", *L'actualité Chimique*, vol. 304, pp. III–XV, 2007.

[FER 08] FERY-FORGUES S., IBANEZ A., DUBERNET B., "Micro et nanocristaux lumineux: ce qui change avec la taille", *L'actualité chimique*, vol. 317, pp. 42–46, 2008.

[FEY 96] FEYMANN R.P., HEY A.J.G, *Feymann Lectures on Computation*, Addison-Wesley, Michigan, 1996.

[FIS 09] FISHLEVICH S., AMIR L., FRIDMAN Y. *et al.*, "Surface display of redox enzymes in microbial fuel cells", *Journal of the American Chemical Society*, vol. 131, pp. 12052–12053, 2009.

[FLE 10] FLEXER V., BRUN N., BACKOV R. *et al.*, "Designing highly efficient enzyme-based carbonaceous foams eletrodes for biofuel cells", *Energy and environment Science*, vol. 3, pp. 1302–1306, 2010.

[FOU 02] FOUASSIER C., "Les luminophores", *L'actualité chimique*, vol. 3, pp. 98–101, 2002.

[GAT 96] GATTESCHI D., CANESCHI A., SESSOLI R., "New magnetic properties emerging from zero and one-dimensional molecular materials", in CORONADO E. *et al.* (eds), *Molecular Magnetism: from Molecular Assemblies to the Devices*, Kluwer Academic Publishers, Dordrecht, 1996.

[GEI 07] GEIM A.K., NOVOSELOV K.S., "The rise of graphene", *Nature Materials*, vol. 6, pp. 183–191, 2007.

[GEO 71] GEORGESCU-ROEGEN N., *The Entropy Law and the Economic Process*, Harvard University Press, Cambridge, 1971.

[GEO 06] GEORGESCU-ROEGEN N., *La décroissance: entropie, écologie, économie*, 3rd edition, Ellébore-sang de la Terre, Paris, 2006.

[GIB 10] GIBSON D.G. *et al.*, "Creation of a bacterial cell controlled by a chemically synthetized genome", *Science*, vol. 329, pp. 52–56, 2010.

[GIL 63] GILMAN J.J., *The Art and Science of Growing Crystals*, John Wiley and Sons, New York, 1963.

[GLA 71] GLANSDORFF P., PRIGOGINE I., *Structure stabilité et fluctuations*, Masson, Paris, 1971.

[GLA 05] GLATTI D.-C., SANQUER M., "La nanoélectronique ne peut être que quantique", *Clefs-CEA Editor*, vol. 52, pp. 14–18, 2005.

[GOG 00] GOGOTSI Y., LIBERS J.A., KALASHNIKOV N. *et al.*, "Graphite polyhedral crystals", *Science*, vol. 290, pp. 317–320, 2000.

[GOG 02] GOGKENHEIMER J.H., OLIVA P.A., "Chaos in Hodgkin-Huxley model", *Journal of Applied Physics Systems*, vol. 1, no. 1, pp. 105–114, 2002.

[GOL 97] GOLDBETER A., *Biochemical Oscillations and Cellular Rhythms: the Molecular Bases of Periodic and Chaotic Behaviour*, Cambridge University Press, Cambridge, 1997.

[GOL 06] GOLDBLATT C., LENTON T.M., WATSON A.J., "Bistability of atmospheric oxygen and the great oscillation", *Nature*, vol. 443, pp. 683–686, 2006.

[GOL 12] GOLDBETER A., GERARD C., GONZE D. *et al.*, "Systems biology of cellular rhythms", *FEBS Letters*, vol. 586, pp. 2955–2965, 2012.

[GON 15] GONNELLA G., MARENDUZZO D., SUMA A. *et al.*, "Motility-induced phase separation and coarsening in active matter", *Comptes Rendus Physique*, Académie des Sciences Paris, vol. 16, pp. 316–331, 2015.

[GOO 13] GOODENOUGH J.B., PARK K.-S., "The Li-ion rechargeable battery: a perspective", *Journal of American Chemical Society*, vol. 136, pp. 1167–1176, 2013.

[GRA 08] GRANDY JR. W.T., *Entropy and the Time Evolution of Macroscopic Systems*, Oxford University Press, New York, 2008.

[GRA 14] GRAETZEL M., "The lightcold shade of perovskite solar cells", *Nature Materials*, vol. 13, no. 9, pp. 838–842, 2014.

[GRE 14] GREEN M.A., HO-BAILLIE A., SNAITH H.J., "The emergence of perovskite solar cells", *Nature Photonics*, vol. 8, pp. 506–514, 2014.

[GRE 15] GREFFET J.-J., BOUCHON P., MARQUIER F., "Dompter les sources incandescentes", *La Recherche* hors-série, vol. 14, pp. 16–19, 2015.

[GRO 09] GROLLIER C., "Comment prévoir l'imprévisible ?", *Pour la Science*, vol. 385, pp. 124–130, 2009.

[GRZ 01] GRZYBOWSKI B.A., WHITESIDES G.M., "Macroscopic synthesis of self-assembled dissipative structures", *Journal of Physical Chemistry B*, vol. 105, pp. 8770–8775, 2001.

[GUE 03] GUEGAN D., *Le chaos en finance : approche statistique*, Economica, Paris, 2003.

[GUI 80] GUINIER A., *La structure de la matière, du ciel bleu à la matière plastique*, Hachette and CNRS, Paris, 1980.

[GUS 09] GUST D., MOORE T.A, MOORE A.L., "Solar fuels via artificial photosynthesis", *Accounts of Chemical Research*, vol. 42, no. 12, pp. 1890–1898, 2009.

[GUT 96] GUTLICH P., JUNG J., GOODWIN H.A., "Spin transitions in Iron (II) complexes" in CORONADO E. *et al.* (eds), *Molecular Magnetism: from Molecular Assemblies to the Devices*, Kluwer Academic Publishers, Dordrecht, 1996.

[HAG 14] HAGIYA M., KONAGAYA A., KOBAYASHI S. *et al.*, "Molecular robots with sensors and intelligence", *Accounts of Chemical Research*, vol. 47, pp. 1681–1690, 2014.

[HAK 77] HAKEN H., *Synergetics*, Springer-Verlag, Berlin, 1977.

[HAN 15] HANNE J., FALK H.J., GORLITZ F. *et al.*, "Stimulated emission depletion nanoscopy with fluorescent quantum dots", *Nature Communication*, vol. 6, no. 7127, 2015.

[HAR 15] HAROCHE S., "Le monde des particules quantiques est subtil et déroutant", *La Recherche*, vol. 501–502, pp. 28–32, 2015.

[HE 09] HE S., CUI Y., YE Y. *et al.*, "Optical nano-antennas and metamaterials", *Materials Today*, vol. 12, no. 12, pp. 16–24, 2009.

[HEE 01] HEEGER A.J., "Nobel lecture: semi-conducting and metallic polymers, the fourth generation of polymeric materials", *Review of Modern Physics*, vol. 73, no. 3, pp. 681–700, 2001.

[HEL 13] HELDER M., CHEN W.-S. *et al.*, "Electricity production with living plants on a green roof: environmental performance of the plant-microbial fuel cell", *Biofuels, Bioproducts and Biorefining*, vol. 7, no. 1, pp. 52–64, 2013.

[HES 71] HESS B., BOITEUX A., "Oscillatory phenomena in biochemistry", *Annual Review of Biochemistry*, vol. 40, pp. 237–258, 1971.

[HIL 11] HILBERT M., LOPEZ P., "The world's technological capacity to store, communicate and compute information", *Science*, vol. 332, pp. 60–65, 2011.

[HON 16] HONG J., LAMBSON B., DHUEY S. *et al.*, "Experimental test of Landauer's principle in single bit operations on nanomagnetic memory bits", *Science Advance*, vol. 2, p. e1501492, 2016.

[HOR 11] HORAJADA P., SERRE C., FERREY G. *et al.*, "Matériaux poreux, stockage et libération de médicaments antitumoraux et antiviraux", *L'actualité chimique*, vol. 348–349, pp. 58–63, 2011.

[HU 14] HU J., LIU S., "Engineering responsive polymer building blocks with host-guest molecular recognition for functional applications", *Accounts of Chemical Research*, vol. 47, pp. 2084–2095, 2014.

HUA 15] HUANG Y., LI F., YE C. *et al.*, "A photochromic sensor microchip for high performance multiplex metal ions detection", *Scientific Reports*, vol. 5, no. 9724, 2015.

[HUB 13] HUBER F., SCHNAUSS J., RONICKE S. *et al.*, "Emergent complexity of the cytoskeleton from single filaments to tissue", *Advances in Physics*, vol. 62, no. 1, pp. 1–112, 2013.

[IMR 99] IMRY Y., LANDAUER R., "Conductance viewed as transmission", *Review of Modern Physics*, vol. 71, pp. S306–S310, 1999.

[INM 13] INMAN M., "The true cost of fossil fuels", *Scientific American*, vol. 308, no. 4, pp. 58–61, 2013.

[IPC 13] IPCC, Assessment report 2013, available at: www.ipcc.ch, 2013.

[ISH 98] ISHIGURO T., YAMAJI K., SAITO G., *Organic Superconductors, Coordinators*, Springer, Berlin, 1998.

[JAF 02] JAFFREZIC-RENAULT N., PIJOLAT C., PAULY A. *et al.*, "Les matériaux pour capteurs chimiques", *L'actualité chimique*, vol. 3, no. 255, pp. 157–171, 2002.

[JAF 09] JAFFREZIC-RENAULT N., "Microcapteurs électrochimiques pour le suivi in situ des contaminants", *L'actualité chimique*, vol. 327–328, pp. 52–55, 2009.

[JAN 17] JANCOVICI J.-M., available at: www.manicore.com/documentation/energie, 2017.

[JAU 01] JAUSSAUD C., JOLY J.-P., MILLION A. *et al.*, "Les modules solaires photovoltaïques: du silicium cristallin aux couches minces", *Clefs CEA*, vol. 44, pp. 29–34, 2001.

[JAY 57] JAYNES E.T., "Information theory and statistical mechanics", *Physical Review*, vol. 106 and 108, pp. 620–630 and 171–190, 1957.

[JAY 03] JAYNES E.T., *Probability Theory: the Logic of Science*, Cambridge University Press, Cambridge, 2003.

[JAY 91] JAYNES E.T., "How should we use entropy in economics?", available at: http://bayes.wutl.edu/etj/node2.htlm, 1991.

[JIA 15] JIA C., PAN F., ZHU Y.G. *et al.*, "High-energy density nonaqueous all redox flow lithium battery enabled with a polymeric membrane", *Sciences Advances*, vol. 1, no. 10, p. e150086, 2015.

[JIN 11] JINEK M., CHYLINSKI K., FONFARA I. *et al.*, "A programmable dual RNA-guidedADN endonuclease in adaptive bacterial immunity", *Science*, vol. 337, pp. 816–821, 2011.

[KAH 88] KAHN O., LAUNAY J.P., "Molecular bistability: an overview", *Chemtronics*, vol. 3, pp. 140–151, 1988.

[KAH 93] KAHN O., *Molecular Magnetism*, VCH Publishing, New York, 1993.

[KEP 14] KEPES F., "La biologie de synthèse, vers une ingénierie du vivant", *Pour la Science*, vol. 440, pp. 28–35, 2014.

[KER 03] KERN H., FAURE P., "Is there is a chaos in the brain? Experimental evidence and related models", *Comptes rendus Biologie*, Académie des Sciences Paris, vol. 326, pp. 787–840, 2003.

[KHO 16] KHORASANINEJAD M., CHEN W.T., DEVLIN R.C. *et al.*, "Metalenses at visible wavelengths: diffraction-limited focusing and subwavelength resolution imaging", *Science*, vol. 352, pp. 1190–1194, 2016.

[KIM 15] KIM Y.D., CHO Y., RYOO J.H., "Bright visible light emission from graphene", *Nature Nanotechnology*, vol. 10, no. 8, pp. 676–681, 2015.

[KIN 14] KINGLER C., "Ces horloges qui rythment notre vie", *La Recherche*, vol. 488, pp. 28–31, 2014.

[KIT 63] KITTEL C., *Quantum Theory of Solids*, 2nd ed., John Wiley and Sons, New York, 1963.

[KIT 04] KITTEL C., *Introduction to Solid State Physics*, 8th ed., John Wiley and Sons, New York, 2004.

[KLE 05] KLEIDON A., FRAEDICH K., "Biotic entropy production and global atmosphere-biosphere interactions", in KLEIDON A., LORENTZ R.D. (eds), *Non Equilibrium Thermodynamics and the Production of Entropy*, Springer-Verlag, Berlin, pp. 173–189, 2005.

[KLI 14] KLINGER C., "Ces horloges qui rythment notre vie", *La Recherche*, vol. 488, pp. 28–31, 2014.

[KON 10] KONDO S., MIURA T., "Reaction-diffusion model as a framework for understanding biological pattern formation", *Science*, vol. 329, pp. 1616–1620, 2010.

[KOS 15] KOSKI J.V., KUTVONEN A., KHAYMOVICH I.M. *et al.*, "On chip Maxwell's demon as an information powered refrigerator", *Physical Review Letters*, vol. 115, p. 260602, 2015.

[KRO 01] KROEMER H., "Nobel lecture: quasielectric band offsets, teaching electrons new tricks", *Review of Modern Physics*, vol. 73, pp. 783–793, 2001.

[KUM 11] KUMMEL R., *The Second Law of Economics: Energy, Entropy and the Origin of Wealth*, Springer, New York, 2011.

[LAN 61] LANDAUER R., "Irreversibility and heat generation in the computing process", *IBM Journal*, vol. 5, no. 3, pp. 183–191, 1961.

[LAS 89] LASKAR J., "A numerical experiment on the chaotic behaviour of the solar system", *Nature*, vol. 338, pp. 237–238, 1989.

[LAU 01] LAUGIER A., DUMON A., "D'Aristote à Mendeleev", *L'actualité Chimique*, vol. 3, no. 244, pp. 38–49, 2001.

[LEB 13] LEBOWITZ J.L., "Boltzmann's entropy and time's arrow", *Physics Today*, vol. 46, pp. 32–38, 1993.

[LEH 10] LEHN J.-M., "Par-delà la synthèse : l'auto-organisation", *Compte-rendus de l'Académie des sciences*, 2010.

[LEM 13] LEMOINE D. *et al.*, "Optical control of an ion channel gate", *Proceedings of the National Academy of Science*, vol. 110, no. 51, pp. 20813–20818, 2013.

[LEM 99] LEMOINE J.-L., *La modélisation des systèmes complexes*, Dunod, Paris, 1999.

[LI 08] LI C., THOSTENSON E.T., CHOU T.-W., "Sensors and actuators based on carbon nanotubes and their composites: a review", *Composites Science and Technology*, vol. 68, pp. 1227–1249, 2008.

[LIB 07] LIBBRECHT K., "La formation des cristaux de neige", *Pour la Science*, vol. 352, pp. 32–39, 2007.

[LIN 16] LIN W., FAN H., WANG Y. *et al.*, "Controlling synchronous patterns in complex networks", *Physical Review E*, vol. 93, p. 042209, 2016.

[LIU 13] LIU Q.-X., DOELMAN A., ROTTCHAFER V. *et al.*, "Phase separation explains a new class of self- organized spatial patterns in ecological systems", *Proceedings of the National Academy of Sciences*, vol. 110, pp. 11905–11910, 2013.

[LIU 15] LIU J., PRINDLE A., HUMPHRIES J. *et al.*, "Metabolic co-dependence gives rise to collective oscillations within biofilms", *Nature,* vol. 523, pp. 550–554, 2015.

[LIU 16] LIU P., JIN Z., KATSUKIS G. *et al.*, "Layered and scrolled nanocomposites with aligned semi-infinite graphene inclusions at the platelet limit", *Science*, vol. 353, no. 6297, pp. 364–367, 2016.

[LIV 11] LIVAGE J., "Vers une chimie douce bio-inspirée", *L'Actualité chimique*, vol. 348–349, pp. 64–69, 2011.

[LOE 07] LOECHLE C., "A 2000-year global temperature reconstruction based on non-treering proxies", *Energy and Environment*, vol. 18, nos. 7–8, pp. 1049–1058, 2007.

[LOG 16] LOGAN B.E., ELIMELECH M., "Membrane-based processes for sustainable power generation using water", *Nature*, vol. 488, pp. 313–319, 2016.

[LOI 06] LOISEAU A., LAUNOIS P., PETIT P. *et al.*, *Understanding Carbon Nanotubes: From Basics to Application*, Springer, Heidelberg, 2006.

[LOK 11] LOKHANDE C.D., DUBAI D.B., JOO O.-S., "Metal oxide based supercapacitors", *Current Applied Physics*, vol. 11, no. 3, pp. 255–270, 2011.

[LU 14] LU Z., MANDAL D., JARZYNSKI C., "Engineering Maxwell's demon", *Physics Today*, vol. 67, p. 60, 2014.

[LUI 13] LUIS F., FERNANDEZ J.F., BARTOLOME J., *Molecular Magnets: Physics and Applications*, Springer, Berlin and Heidelberg, 2013.

[MAL 15] MALYSHEV D., ROMESBERG F.E., "The expanded genetic alphabet", *Angewandte Chemie International Edition*, vol. 54, pp. 11930–11944, 2015.

[MAN 74] MANDELBROT B., *Les objets fractals: forme, hasard et dimension*, Champ Sciences, Flammarion, Paris, 1974.

[MAN 14] MANDELBROT B., *Fractales, hasard et finances*, Flammarion, Paris, 2014.

[MAR 06] MARTYUSHEV L.M., SELEZNEV V.D., "Maximum entropy production in physics, chemistry and biology", *Physics Reports*, vol. 42, pp. 1–45, 2006.

[MAR 95] MARCHAND A., PACAULT A., *La thermodynamique mot à mot*, De Boeck-Wesmael, Brussels, 1995.

[MAR 08] MARINESCO S., PERNOT P., "Biocapteurs implantables in vivo", *Techniques de l'Ingénieur RE*, vol. 108, pp. 1–13, 2008.

[MAR 09] MARONEY O., "Information processing and thermodynamic entropy", in ZALTA E.N. (ed.), *The Standford Encyclopedia of Philosophy*, available at: http://plato.standford.edu/archives, 2009.

[MAT 80] MATISOO J., "Overview of Josephson technology logic and memory", *IBM Journal of Research and Development*, vol. 24, no. 2, pp. 113–129, 1980.

[MEI 82] MEINHARDT H., *Models of Biological Pattern Formation*, Academic Press, London 1982.

[MEZ 90] MÉZARD M., NADAL J.P., "Réseaux de neurones et physique statistique", *Images de la Physique éditions*, CNRS, pp. 99–105, 1990.

[MIL 07] MILES R.W., ZOLPI G., FORBES I., "Inorganic photovoltaïc cells", *Materials Today*, vol. 10, pp. 20–27, 2007.

[MIN 00] MINGOTAUD C., DELHAES P., MEISEL M.W. *et al.*, "Magnetic Langmuir-Blodgett films", in MILLER J.S., DRILLON M. (eds), *Magnetism: Molecules to Materials II*, Wiley-VCH Ed., New York, 2000.

[MIN 15] MINEV I.V., MUSIENKO P., HIRSCH A., "Electronic dura-mater for long term multimodal neural interfaces", *Science*, vol. 347, no. 6218, pp. 159–163, 2015.

[MIO 15] MIOMANDRE F., AUDEBERT P., "Couplage électrochimie-luminescence: développement instrumental et systèmes électrofluorochromes", *L'actualité chimique*, vol. 400–401, pp. 12–16, 2015.

[MIR 02] MIR S., *Dispositifs et physique des microsystèmes sur Silicium*, Hermès Science Lavoisier, Paris, 2002.

[MO 16] MO R., GU Z., "Tumor microenvironment and intracellular signal-activated nanomaterials for anticancer drug delivery", *Materials Today*, vol. 19, no. 5, pp. 274–283, 2016.

[MOE 12] MOERNER W.E., "Microscopy beyond the diffraction limit using actively controlled single molecules", *Journal of Microscopy*, vol. 246, pp. 213–220, 2012.

[MOI 09] MOISAN S., POULIN P., DERRE A., "Les fibres nanocomposites de nanotubes de carbone", *L'actualité chimique*, vol. 336, pp. 31–35, 2009.

[MON 70] MONOD J., *Le hasard et la nécessité, essai sur la philosophie naturelle de la biologie moderne*, Le Seuil, Paris, 1970.

[MON 16] MONROE C., SCHOELKOPF R., LUKIN M., "L'ordinateur quantique en kit", *Dossier pour la science*, vol. 93, pp. 62–68, 2016.

[MOR 14] MORIN S. A., SHEVCHENKO Y., LESSING J. *et al.*, "Using 'click-e-bricks to make 3D Elastomeric structures", *Advanced Materials*, vol. 26, pp. 5991–5999, 2014.

[MOR 90] MORIN E., *Introduction à la pensée complexe*, Le Seuil, Paris, 1990.

[MOU 15] MOULIN E., FUFS G., GAVAT O. *et al.*, "Systèmes chimiques auto-organisés et matériaux", *L'actualité chimique*, vol. 399, pp. 25–30, 2015.

[MUE 10] MUELLER T., KINOSHITA M., STEINER M. *et al.*, "Efficient narrow-band light emission from a single carbon nanotube p-n diode", *Nature Nanotechnology*, vol. 5, pp. 27–31, 2010.

[MUL 12] MULLER P., ROGERS K.W., JORDAN B.M. *et al.*, "Differential diffusivity of nodal and lefty underlies a reaction-diffusion patterning system", *Science*, vol. 336, no. 6082, pp. 721–724, 2012.

[MUL 69] MULLIKEN R.S., PERSON W.B., *Molecular Complexes*, John Wiley and Sons, New York, 1969.

[NAK 94] NAKAMURA S., MUKAI T., SEOH M., "Candela class high-brightness INGaN/AlGaN double heterostructure blue light emitting diodes", *Applied Physics Letters*, vol. 64, pp. 1687–1689, 1994.

[NAK 08] NAKATANI K., MICHEAU J.-C., "La photocommutation: des verres photochromiques aux machines moléculaires", *L'actualité chimique*, vol. 317, p. 3137, 2008.

[NI 15] NI X. *et al.*, "An ultrathin invisibility skin cloak for visible light", *Science*, vol. 349, no. 6254, pp. 1310–1314, 2015.

[NOO 13] NOORDUIN W.L., GRINTHAL A., MAHADEVAN L. *et al.*, "Rationally designed complex hierarchical microarchitectures", *Science*, vol. 430, pp. 832–837, 2013.

[NOR 14] NORVIG P. CHABALIER M., "L'intelligence artificielle", *Le monde des Sciences*, vol. 3, pp. 47–54, 2014.

[NYE 61] NYE J.F., *Propriétés physiques des cristaux, leur représentation par des tenseurs et des matrices*, Dunod, Paris, 1961.

[ODR 13] ODRU P., *Le stockage de l'énergie*, 2nd ed., Dunod, Paris, 2013.

[OFT 88] OFTA, *L'électronique moléculaire: perspectives en matière de traitement moléculaire de l'information*, Masson, Paris, 1988.

[PAD 13] PADIRAC D., "Un système artificiel imite la dynamique des écosystèmes", *La Recherche*, vol. 482, pp. 59–61, 2013.

[PAR 15] PARK J., KIM M., LEE Y. *et al.*, "Fingertip skin-inspired microstructured ferroelectric skins discriminate static/dynamic pressure and temperature stimuli", *Science Advances*, vol. 1, no. 9, p. e1500661, 2015.

[PAR 16] PARK S.-J. *et al.*, "Phototactic guidance of a tissue-engineered soft-robotic ray", *Science*, vol. 353, no. 6295, pp. 158–162, 2016.

[PAS 96] PASSET R., *L'économique et le vivant*, 2nd ed., Economica, Paris, 1996.

[PAS 14] PASTORE A., PIONTI Y., FERREIRA DA COSTA GOMES M. *et al.*, "The infection of global epidemies", *Network Science*, vol. 2, pp. 132–137, 2014.

[PAT 96] PATRICK J. W., *Porosity in Carbons*, Edward Arnold, London, 1996.

[PEL 00] PELCE P., *Théorie des formes de croissance: digitations, dendrites et flammes*, Editions EDP sciences et CNRS, Paris, 2000.

[PEN 14] PENG L.M., ZHANG Z., WANG S.,"Carbon nanotube electronics: recent advances", *Materials Today*, vol. 17, no. 9, pp. 433–442, 2014.

[PER 10] PERROT A., NGUYEN D., BOURGEAT-LAMI E. *et al.*, "Les particules mettent les formes: des atomes aux molécules colloïdales", *L'actualité chimique*, vol. 340, pp. 18–21, 2010.

[PER 13a] PERERA U.G.E., AMPLE F., KERSELL H. *et al.*, "Controlled clockwise and anticlockwise rotational switching of a molecular motor", *Nature Nanotechnology*, vol. 8, pp. 46–51, 2013.

[PER 13b] PERETO J., CATALA J., MORENO A. *et al.*, "Un système autonome et évolutif", *Les dossiers de la Recherche*, vol. 2, pp. 18–22, 2013.

[PER 13c] PERNIOLA L., "Les mémoires du futur", *Pour la Science*, vol. 433, pp. 92–99, 2013.

[PET 16] PETERSON J.P.S., SARTHOUR R.S., SOUZA A.M. *et al.*, "Experimental demonstration of information to energy conversion in a quantum system at the Landauer limit", *Proceedings of the Royal Society A*, vol. 472, pp. 21–88, 2016.

[PFA 14] PFAFF W., HENSEN B., BERNIEN H. *et al.*, "Unconditional quantum teleportation between distant solid-state quantum bits", *Science-express*, vol. 101126, pp. 532–535, 2014.

[POC 02] POCHOLLE J.-P., VIVIEN D., "Les matériaux laser pour l'optique non linéaire", *L'actualité chimique*, vol. 3, pp. 116–128, 2002.

[PRE 13] PRESSÉ S., GHOSH K., LEE J. *et al.*,"Principles of maximum entropy and maximum caliber in statistical Physics", *Review of Modern Physics*, vol. 85, pp. 1115–1141, 2013.

[PRI 68] PRIGOGINE I., *Introduction à la thermodynamique des processus irréversibles*, Monographies Dunod, Paris, 1968.

[PRI 72] PRIGOGINE I., NICOLIS G., BABLOYANTZ A., "Thermodynamics of evolution", *Physics Today*, vol. 25, no. 11, pp. 23–28, 1972.

[PRI 94] PRIGOGINE I., *Les lois du chaos*, Flammarion, Paris, 1994.

[PRO 08] PROCHIANTZ A., *Géométries du vivant*, Fayard, Paris, 2008.

[PUE 14] PUETZ S.J., PROKOPH A., BORCHARDT G. *et al.*, "Evidence of synchronous, decadal to billion year cycles in geological, genetic and astronomical events", *Chaos, Solitons and Fractals*, vols. 62–63, pp. 55–75, 2014.

[RAB 89] RABE J.P., "Imaging surfaces, electric charges and magnetic domains with the atomic force microscope", *Angewandte Chemical International Edition*, vol. 28, no. 11, pp. 1578–1581, 1989.

[RAM 03] RAMADE F., *Eléments d'écologie, écologie fondamentale*, 3rd edition, Dunod, Paris, 2003.

[RAS 14] RASPOPOVIC J., MARCON L., RUSSO L. *et al.*, "Digit pattern is controlled by a Bmp-Sox9-Wnt Turing network modulated by morphogen gradients", *Science*, vol. 345, pp. 566–570, 2014.

[REE 11] REECE S.Y., HAMEL J.A., SUNG K. *et al.*, "Wireless solar water splitting using Silicon-based semiconductors and earth abundant catalysts", *Science*, vol. 334, pp. 645–648, 2011.

[RIC 99] RICARD J., *Biological Complexity and the Dynamics of Life Processes*, Elsevier, Amsterdam, 1999.

[RIF 12] RIFKIN F., *La troisième révolution industrielle : comment le pouvoir latéral va transformer l'énergie, l'économie et le monde*, Les liens qui libèrent, Paris, 2012.

[ROB 05] ROBLEDO A., "Critical attractors and q-statistics", *Europhysics News*, vol. 36, no. 6, pp. 214–218, 2005.

[ROM 14] ROMNEY B., "L'ordinateur quantique: 3. Construire le qubit parfait", *La Recherche*, vol. 485, pp. 36–40, 2014.

[ROS 79] ROSMORDUC J., *De Thalès à Einstein*, Etudes vivantes, Paris, 1979.

[ROT 16] ROTH W.J., GIL B., MAKOWSKI W. *et al.*, "Layer like porous materials with hierarchical structure", *Chemical Society Review*, vol. 45, no. 12, pp. 3400–3438, 2016.

[ROU 09] ROUQUEYROL F., LUCIANI L., LLEWELLYN P. *et al.*, "Texture des matériaux pulvérulents et poreux", *Techniques de l'ingénieur, analyse et caractérisation*, vol. P-1050, pp. 1–24, 2009.

[RUD 11] RUDERNAC T., RUANGSUPAPICHAT H., MACIA B. *et al.*, "Electrically driven directional motion of a four-wheeled molecule on a metal surface", *Nature,* vol. 479, pp. 208–211, 2011.

[RUE 82] RUELLE D., "Do turbulent crystals exist?", *Physica*, vol. 113A, pp. 619–623, 1982.

[RUE 92] RUELLE D., "Déterminisme et prédicibilité", in DE GENNES P.-G. (ed.), *L'ordre du chaos*, Belin, Paris, 1992.

[RUT 05a] RUTH M., "Insights from thermodynamics for the analysis of economic processes", in KLEIDON A., LORENTZ R.D. (eds), *Non-Equilibrium Thermodynamics and the Production of Entropy*, Springer-Verlag, Berlin, 2005.

[RUT 05b] RUTHERFORD A.W., "De l'hydrogène à partir du soleil et de l'eau", *Clefs CEA*, vol. 50–51, pp. 49–51, 2005.

[SAD 92] SADOC J.F., MOSSIERI R., "Un ordre caché dans la matière désordonnée", in GENNES P.-G. (ed.), *L'ordre du chaos*, Paris, 1992.

[SAL 14] SALANNE M., ROTEMBERG B., SIMON P., "Plus d'électricité dans le carbone", *La Recherche*, vol. 491, pp. 52–55, 2014.

[SAL 17] SALEH M.S., HU C., PANAT R., "Three-dimensional microarchitected materials and devices using nanoparticle assembly by pointwise spatial printing", *Science Advances*, vol. 3, no. 3, p. e1601986, 2017.

[SAM 13] SAMPALO J., CROS V., ROHART S. *et al.*, "Nucleation, stability and current-induced motion of isolated magnetic skyrmions in nanostructures", *Nature Nanotechnology*, vol. 8, pp. 839–844, 2013.

[SAN 05] SANDS D., *Diode Lasers*, Institute of Physics Publishing, Bristol and Philadelphia, 2005.

[SAO 13] SAOMORI P., "Nanochemistry: from single molecules to functional supramolecular materials and devices", *L'actualité chimique*, vol. 380, pp. III–X, 2013.

[SAU 16] SAUVAGE J.-P., DUPLAN V., NIESS F., "Systèmes moléculaires contractiles et extensibles: vers des muscles artificiels", *L'actualité chimique*, vol. 406, pp. 13–20, 2016.

[SCH 39] SCHUMPETER J.A., *Business Cycles: a Theoretical, Historical and Statistical Analysis of the Capitalist Process*", McGraw Hill, New York, 1939.

[SCH 44] SCHRODINGER E., *What is Life?*, Berkeley University Press, San Francisco, 1944.

[SCH 05] SCHUSTER H.-G., JUST W., *Deterministic Chaos: an Introduction*, 4th ed., John Wiley and Sons, New York, 2005.

[SCH 10] SCHUMACHER B., WESTMORELAND M., *Quantum Theory Processes Systems and Information*, Cambridge University Press, Cambridge, 2010.

[SCH 14] SCHEFFE J.R., STEINFELD A., "Oxygen exchange materials for solar thermochemical spitting of H_2O and CO_2: a review", *Materials Today*, vol. 17, no. 7, pp. 341–348, 2014.

[SEG 13] SEGAL J., "La théorie de l'information existe-t-elle?", *Pour la Science*, vol. 424, pp. 19–25, 2013.

[SEI 12] SEIFFERT U., "Stochastic thermodynamics, fluctuation theorems and molecular machine", *Reports Progress on Physics*, vol. 75, no. 12, p. 126001, 2012.

[SEL 16] SELOSSE M.-A., "Au-delà de l'organisme l'holobionte", *Pour la Science*, vol. 469, pp. 80–84, 2016.

[SEM 16] SEMENOV S.N., KRAFT L.J., AINIA A. *et al.*, "Autocatalytic, bistable, oscillatory networks of biologically relevant organic reactions", *Nature*, vol. 537, no. 7622, pp. 656–660, 2016.

[SEO 10] SEO D., SONG H., "Synthesis of gold particles in liquid phase", in LOUIS C., PLUCHERY O. (eds), *Gold Nanoparticles*, Imperial College London, London, 2010.

[SER 13] SERVICE R.F., "Turning up the light", *Science*, vol. 342, pp. 794–797, 2013.

[SET 16] SETZIER B.P., ZHUANG Z., WITTKOPF J.A. *et al.*, "Activity targets for nanostructured platinium-group-metal free catalysts in hydroxide exchange membrane fuel cells", *Nature Nanotechnology*, vol. 11, pp. 1020–1025, 2016.

[SIL 02] SILLON B., "Les polymères à mémoire de forme", *L'actualité chimique*, vol. 3, pp. 182–188, 2002.

[SIM 58] SIMONDON G., *Du mode d'existence des objets techniques*, Aubier, Paris, 1958.

[SIM 09] SIMON P., TARASCON J.-M., "Stockage électrochimique de l'énergie", *L'actualité chimique*, vol. 327–328, pp. 87–96, 2009.

[SIR 13] SIRIA A., BIANCE A.-L., BOCQUET L., "Une énergie en osmose avec l'avenir", *La Recherche*, vol. 482, pp. 50–53, 2013.

[SNY 08] SNYDER G.J., TOBERER E.S., "Complex thermoelectric materials", *Nature Materials*, vol. 7, pp. 105–114, 2008.

[SOK 11] SOKOLOV V.V., ZHIROV O.V., KHARKOV Y.A., "Classical versus quantum dynamical chaos: sensitivity to external perturbations, stability and reversibility" in *Chaos Theory: Modeling, Simulation and Applications*, World Scientific Editors, Singapore, 2011.

[STO 09] STODDART J.F., "The master of chemical topology", *Chemical Society Review*, vol. 38, pp. 1521–1529, 2009.

[STR 08] STRUKOV D.B., SNIDER G.S, STEWART D.R. *et al.*, "The missing memristor found", *Nature*, vol. 453, pp. 80–83, 2008.

[TAN 87] TANG C.W., VAN SLYKE S.A., "Organic electroluminescent diodes", *Applied Physics Letters*, vol. 51, pp. 913–917, 1987.

[THO 72] THOM R., *Stabilité structurelle et morphogénèse*, Benjamin, Paris, 1972.

[THO 14] THONDAMAI M., WITTING M., SCHMITT-KOPPLIN P. *et al.*, "Steroid hormone signaling links reproduction to lifespan in dietary-restricted Caenorhabditis elegans", *Nature Communications*, vol. 5, 2014.

[TIN 13] TINGRY S., CRETIN M., INNOCENT I., "Les biopiles enzymatiques pour produire de l'électricité", *L'actualité chimique*, vol. 373, pp. 18–25, 2013.

[TOY 10] TOYABE S., SAGAWA T., UEDA M. *et al.*, "Experimental demonstration of information to energy conversion and validation of the generalized Jarzynski equality", *Nature Physics*, vol. 6, pp. 988–992, 2010.

[TRE 09] TREMILLION B., "L'évolution de l'électrochimie", *L'actualité chimique*, vol. 327–328, pp. 10–26, 2009.

[TRE 15] TREMBLAY J.P., "CRISPR un système qui permet de corriger ou de modifier l'expression de gènes responsables de maladies héréditaires", *Médecine-Sciences*, vol. 31, no. 11, pp. 1014–1022, 2015.

[TSA 11] TSALLIS C., "The nonadditive entropy Sq and its applications in physics and elsewhere", *Entropy*, vol. 13, pp. 1765–1804, 2011.

[TUR 52] TURING A., "The chemical basis of morphogenesis", *Philosophical transactions of the Royal Society*, vol. 327, pp. 37–42, 1952.

[TZE 17] TZEDAKIS P.C., CRUCIFIX M., MITSUI T. *et al.*, "A simple rule to determine which insolation cycles lead to interglacials", *Nature*, vol. 542, pp. 427–432, 2017.

[ULM 91] ULMAN A., *An Introduction to Ultrathin Organic Films from Langmuir-Blodgett to Self-Assembly*, Academic Press, New York, 1991.

[USS 10] US SECURITIES AND EXCHANGE COMMISSION, Finding regarding the market events of May 6th 2010, SEC and CFTC Report, 2010.

[UPT 80] UPTON F.P., "Electricity as power", *Science*, vol. 1, p. 5, 1880.

[VEL 11] VELASCO R.M., SCHERER GARCIA-COLIN L., URIBE F.J. *et al.*, "Entropy production: its role in non-equilibrium thermodynamics", *Entropy*, vol. 13, pp. 82–116, 2011.

[VEN 84] VENABLES J.A., SPILLER G.D.T., HANBUCKEN M., "Nucleation and growth of thin films", *Report Progress in Physics*, vol. 47, pp. 399-459, 1984.

[VER 01] VERDAGUER M., "La molécule, précurseur du solide magnétique", *Images de la Physique 2000*, pp. 81–87, CNRS, Paris, 2001.

[VID 88] VIDAL C., LEMARCHAND A., *La réaction créatrice*, Hermann, Paris, 1988.

[VID 94] VIDAL C., DEWEL G., BOCKMANS P., *Au-delà de l'équilibre*, Hermann, Paris, 1994.

[VIL 12] VILLANI C., *Théorème vivant*, Grasset et l'aquarelle, Paris, 2012.

[VIL 17] VILLIEU P., "Monnaie: théorie économique de la monnaie", *Encyclopédie Universalis*, available online, 2017.

[VIN 02] VINCENT J.F.V., "Survival of the cheapest", *Materials Today*, vol. 28, pp. 414–420, 2002.

[VIN 12] VINCENT R., KLYATSKAYA S., RUBEN M. *et al.*, "Electronic read-out of a single nuclear spin using a molecular spin transistor", *Nature*, vol. 488, pp. 357–360, 2012.

[WAS 10] WASSEL J.K., KANER R.B., "Graphene, a promising transparent conductor", *Materials Today*, vol. 13, no. 3, pp. 52–59, 2010.

[WAU 14] WAUTELET M. *et al.*, *Les Nanotechnologies*, 3rd ed., Dunod, Paris, 2014.

[WEI 48] WIENER N., *Cybernetics or Control and Communication in the Animal and the Machine*, John Wiley and Sons/Hermann and Company, New York/Paris 1948.

[WEI 16] WEILENMANN M., KRAEMER L., FAIST P. *et al.*, "Axiomatic relation between thermodynamic and information-theoric entropies", *Physical Review Letters*, vol. 117, p. 260601, 2016.

[WHI 99] WHITESIDES G.M., ISMAGILOV R.F., "Complexity in chemistry", *Science*, vol. 284, pp. 89–92, 1999.

[WHI 01] WHITE S.R., SOTTOS N.R., GEUBELLE P.H. *et al.*, "Autonomic healing of polymer composites", *Nature*, vol. 409, pp. 794–797, 2001.

[WIF 10] WIFFELS R.H., BARBOSA M.J., "An outlook on microalgal biofuels", *Science*, vol. 329, pp. 796–799, 2010.

[WIK 17] WIKIPEDIA, "Business cycles", available online, 2017.

[WOL 01] WOLFE J., *Cellular Thermodynamics*, John Wiley and Sons, New York, 2001.

[XU 15] XU B., JACQUIR S., BINCZAK S., YAHIA H. *et al.*, "Classification of cardiac arrhythmia in vitro based on multivariate complexity analysis", www.cinc2015.org/>.<hal-01163088v2>, 2015.

[YUA 83] YUAN J.-M., LIU E., TUNG M., "Bistability and hysteresis in Laser-driven polyatomic molecules", *Journal of Chemical Physics*, vol. 79, no. 10, pp. 5034–5040, 1983.

[ZHA 12] ZHANG F., XU L., ZHANG K. *et al.*, "The potential and flux landscape theory of evolution", *Journal of Chemical Physics*, vol. 137, no. 065102, 2012.

[ZYS 91] ZYSS J., "Origin of microscopic nonlinearity in organic systems", in PRASAD P.N., WILLIAMS D.J. (eds), *Introduction to Nonlinear Optical Effects in Molecules and Polymers*, Academic Press, New York, 1991.

Index

N, O

P

Q, R

S

T

V, W, Z

Other titles from

ISTE

in

Materials Science

2017

CARVELLI Valter, JAIN Atul, LOMOV Stepan
Fatigue of Textile and Short Fiber Reinforced Composites
(Durability and Ageing of Organic Composite Materials Set – Volume 1)

2016

TORRENTI Jean-Michel
Materials and Infrastructures 1
Materials and Infrastructures 2

RÉMOND Yves, AHZI Saïd, BANIASSADI Majid, GARMESTANI Hamid
Applied RVE Reconstruction and Homogenization of Heterogeneous Materials

2015

JOUD Jean-Charles, BARTHES-LABROUSSE Marie-Geneviève
Physical Chemistry and Acid-Base Properties of Surfaces

LALEVEE Jacques, FOUASSIER Jean-Pierre
Dyes and Chromophores in Polymer Science

2014

HAMAIDE Thierry, DETERRE Rémi, FELLER Jean-François
Environmental Impact of Polymers

LAFLEUR Pierre, VERGNES Bruno
Polymer Extrusion

2013

BATHIAS Claude
Fatigue Limit in Metals

LEXCELLENT Christian
Shape-memory Alloys Handbook

2012

DAVIM J. Paulo
Wear of Advanced Materials

DELHAES Pierre
Carbon Science and Technology: From Energy to Materials

GOUDEAU Philippe, GUINEBRETIERE René
X-Rays and Materials

SAANOUNI Khemais
Damage Mechanics in Metal Forming: Advanced Modeling and Numerical Simulation

VANBÉSIEN Olivier
Artificial Materials

VERDU Jacques
Oxidative Ageing of Polymers

2011

BATHIAS Claude, PINEAU André
Fatigue of Materials and Structures: Application to Damage

BATHIAS Claude, PINEAU André
Fatigue of Materials and Structures: Application to Design

PRIESTER Louisette
Grain Boundaries and Crystalline Plasticity

VIGNES Alain
Extractive Metallurgy 1: Basic Thermodynamics and Kinetics
Extractive Metallurgy 2: Metallurgical Reaction Processes
Extractive Metallurgy 3: Processing Operations and Routes

2010

BATHIAS Claude, PINEAU André
Fatigue of Materials and Structures: Fundamentals

CHATEIGNER Daniel
Combined Analysis

CHEVALIER Yvon, VINH TUONG Jean
Mechanical Characterization of Materials and Wave Dispersion V1
Mechanical Characterization of Materials and Wave Dispersion V2

DELHAES Pierre
Carbon-based Solids and Materials

2009

ALVAREZ-ARMAS Iris, DEGALLAIX-MOREUIL Suzanne
Duplex Stainless Steels

DAVIM J. Paulo
Machining Composite Materials

GALERIE Alain
Vapor Surface Treatments

2008

BATTAGLIA Jean-Luc
Heat Transfer in Materials Forming Processes

BLONDEAU Régis
Metallurgy and Mechanics of Welding

FRANÇOIS Dominique
Structural Components

REYNE Maurice
Plastic Forming Processes

TAKADOUM Jamal
Materials and Surface Engineering in Tribology

2007

BOCH Philippe, NIEPCE Jean-Claude
Ceramic Materials

CRISTESCU Constantin
Materials with Rheological Properties

2006

LASSEN Tom, RÉCHO Naman
Fatigue Life Analyses of Welded Structures

Printed and bound by CPI Group (UK) Ltd, Croydon, CR0 4YY
13/09/2026
14961039-0006